THE BREACHER'S PLAYBOOK

A DEKLAN NOVAK NOVEL

NAVY SEAL TOM HRUBY AND JAMES POMERANTZ

CONTENTS

The Deklan Novak Timeline:	ix
Prologue	1
Chapter 1	3
Chapter 2	17
Chapter 3	25
Chapter 4	47
Chapter 5	59
Chapter 6	69
Chapter 7	79
Chapter 8	85
Chapter 9	93
Chapter 10	111
Chapter 11	117
Chapter 12	127
Chapter 13	145
Chapter 14	153
Chapter 15	167
Chapter 16	179
Chapter 17	187
Chapter 18	195
Chapter 19	205
Chapter 20	209
Chapter 21	227
Chapter 22	235
Chapter 23	247
Chapter 24	253
Chapter 25	263
Chapter 26	271
Chapter 27	279
Chapter 28	287
Chapter 29	297
Chapter 30	301
Chapter 31	309
Chapter 32	317
Chapter 33	327

About the Authors	339
Acknowledgments	343

THE BREACHER'S PLAYBOOK

A Deklan Novak Novel

Copyright © 2024 by Tom Hruby and James Pomerantz

All rights reserved. No part of this publication may be reproduced, distributed, or transmitted in any form or by any means (including photocopying, recording, or other electronic or mechanical methods), without the prior written permission of the publisher, except in the case of brief quotations in a book review and certain other noncommercial uses permitted by copyright law.

First Edition

The opinions and or/views which are expressed in this work are solely those of the author and do not necessarily reflect the views or opinions of the publisher. Their appearance in this publication does not constitute an endorsement by Tactical 16 Publishing, its affiliates, or its employees. The contents and information conveyed herein is based upon information that the author considers reliable, but neither its completeness or accuracy are warrantied by the publisher, and it should not be relied upon as such. Tactical 16 Publishing hereby disclaims any responsibility or liability to any party for the contents of this publication.

This is a work of fiction. Names, characters, places, and events are either the product of the author's imagination or are used fictitiously. Any resemblance to actual persons (living or dead), events, or locations is entirely coincidental. No references made are intended to represent (and neither should they be inferred to represent) reality.

Published by Tactical 16 Publishing

Colorado Springs, Colorado

www.Tactical16.com

ISBN: 978-1-966413-00-4 (paperback)

978-1-943226-99-3 (hard back)

PRAISE FOR THE BREACHER'S PLAYBOOK

"Highly Efficient Operator and Tactician. Effective and Trustworthy Leader. Eerily Prophetic, a Truly Divergent Thinker. Seemingly Limitless, Enigmatic, and nearly Impossible to Understand. It is Terrifying to Imagine a Mind and Body like Tom not on Our Side. God Bless America. God Bless Freedom."

ANONYMOUS SEAL TEAM 6 RED SQUADRON MEMBER AND CIA OPERATIVE

"As most SEALs frown upon the gluttony of books recanting the often true and many times, enhanced accounts of their Special Forces brethren, Tom Hruby's story is as remarkable and unique as they get. A true warrior in every sense. *The Breacher's Playbook* is not about ego. The book is all about determination at the highest level and the inspirational message that anything is possible."

SEAL TEAMMATE OF TOM HRUBY, BUD/S CLASS 264/266

"As a former NFL player, the unprecedented discipline within a SEAL team has always been inspiring. So, after serving more than a decade as a Navy SEAL and retired with a decorated combat career, Deklan Novak realized his days of making life and death decisions were not behind him. *The Breacher's Playbook* is a book I simply could not put down."

JIM MORRISSEY - LINEBACKER CHICAGO BEARS 1985-1993 WORLD CHAMPION CHICAGO BEARS 1985 SUPER BOWL XX

Deklan: (Deklan or Deklan is short for the Biblical name MelchizeDeklan: King of Salem, King of Righteousness, High Priest of the brotherhood of Aaron and Christ...*And MelchizeDeklan king of Salem brought out bread and wine: and he was [is] the priest of the most high God. And he blessed him, and said, 'Blessed be Abram to the most high God, possessor of heaven and earth, And blessed be the most high God, which hath delivered thine enemies into thy hand'. And he gave him tithe from all.*)
—*Genesis 14:18–20, King James Version*[14]

THE DEKLAN NOVAK TIMELINE:

All-Conference high school linebacker and wrestler in suburban Chicago. 1997-2000

Promising 18-year-old MMA professional fighter at 7-0, forgoing college. 2001

9/11 attacks on New York City changed lives in profound ways. 9/11/2001

Navy boot camp at Recruit Training Command, Great Lakes (RTC Great Lakes) 2006   North Chicago, Illinois.

BUD/S Class 264 & 266...a Navy SEAL Trident 2007-2008: Coronado, California.

Assigned to SEAL Team One based in Coronado, California. 2008

DEPLOYMENTS:
SEAL Team One (ST1) @ FOB Chapman & FOB Salerno...Khoust, Afghanistan 2009,

Philippines 2010,

ST1 @ Kirkuk, Iraq 2011, ST1 @ Tikrit, Iraq 2011,

Lebanon 2012.

AWARDS:

Joint Service Achievement Medal

Navy and Marine Corp Achievement Medal (2)

Navy Good Conduct Medal (2)

Afghanistan Campaign Medal

Global War on Terrorism Expeditionary Medal

Sea Service Deployment Medal

Navy and Marine Corp Overseas Service Ribbon

BUD/S Prep Instructor at Naval Station (NAVSTA) Great Lakes 2012   North Chicago, Illinois

Full-time student at Northwestern University while an active-duty Navy SEAL instructor at NAVSTA Great Lakes.

Linebacker at Northwestern University 2013-2016 @ 32, 33, 34 years-old

Official retirement from the United States Navy 2015

2014 win over Notre Dame in South Bend, Outback Bowl 2015, Pinstripe Bowl 2016

A degree in Criminal Justice from Northwestern University 2016

The FBI Special Agent Training Academy in Quantico, Virginia 2018

The FBI Violent Crimes Against Children Squad (VCAC) 2019

> We are getting way ahead of ourselves...let's first meet Deklan (Deklan) Novak.

PROLOGUE

FBI SPECIAL AGENT and former Navy SEAL Deklan Novak wrote recently about the ongoing war on terror:

"Every expression of legally justified violence is controlled. Legally justified violence is always in response to uncontrolled violence, the last refuge of incompetence according to a Boston University professor named Isaac Asimov. The rules of engagement in sports, war and law enforcement are bound by moral, statutory, and ethical restraints. The rules of engagement in war and law enforcement are in place to combat barbarism, yet the protectors are forced to wallow in the mire. I was told repeatedly as a child and as a young man that my violent nature was inevitably charting my future for a predictable degenerate life in prison and an early abbreviated eulogy. Sports and the military were prescribed alternatives to a criminal destiny. In the military, I trained for years to face down the uncontrolled global pandemic of terrorism. Our efforts were shut down at every level when the beast was found. No clearer example can be found than the aftermath of the October 7, 2023, unprovoked attacks on Israeli civilians. In football, violence is used to promote the product, yet penalized and ejected when practiced. The old version of insanity was doing the same thing repeatedly and expecting different results. The newer version of insanity is to combat uncontrolled brutality with fear and restraint to induce civility. Detail the time in our history where controlled, regulated, and ethical restraint has eradicated what the presence of God has been unable to do since man arrived on the planet. God said…***Be fruitful and multiply, behold children are a heritage from the Lord. Like arrows in the hand of a warrior.*** If God bears no culpability when those we love more than life itself are lost to evil men, then judge not the means to remove the evil."

CHAPTER ONE

2013 CAMP KENOSHA...THE UNIVERSITY OF WISCONSIN – PARKSIDE CAMPUS & NORTHWESTERN UNIVERSITY SUMMER FOOTBALL TRAINING CAMP

THE NONDESCRIPT NAVY duffle bag was tossed onto the nearest bed in the dorm suite. Northwestern University held their summer football training camp each year on the campus of the University of Wisconsin-Parkside (UWP) in Kenosha, Wisconsin, some 55 miles north of Evanston, Illinois. Kenosha is the fourth largest city in Wisconsin located on the southwestern shore of Lake Michigan. Kenosha is a satellite city within the greater Chicago metropolitan area. The campus is a tree-lined Rockwell painting comprised of just under 5,000 students amidst 15 Gothic and Georgian style buildings surrounded by restored prairies, oak & maple forests, and meandering creeks.

Novak surveyed the dorm room he had been assigned with three other linebackers. Roger Dixon, Peter Branch and Marty Reyes were at the team meeting that Novak was expected to attend. The UWP dorm room had a plethora of electronic gear set-up like a Best Buy showroom. There were two flat screen televisions, video game consoles and a Harmon Kardon high-performance audio/video receiver that resembled a recording studio. The receiver was a seven-channel surround sound receiver with HDMI 2.0 upscaling to 1080p. The AVR boosted 40 watts on five channels to drive the home theater speakers with authority. Jesus Christ! Novak dabbled in Call of Duty on various bases and some SEALs acted as consultants for Treyarch, but gamers were not generally in Deklan's wheelhouse for running mates. Novak brought some gym clothes, one pair of jeans, tee shirts, running shoes and his Navy skivvies. Petty Officer First Class Deklan Novak hadn't played football for almost fifteen years. Novak's last game was the playoff loss during the Illinois State High School 7A tournament in 1999. Deklan was one of 27 players in 2013 invited to Northwestern University's summer football camp as walk-on players. The Big Ten allowed 112 roster spots that included 85 scholarship players. At Northwestern, the walk-on roster spots were predetermined due to the difficult admissions standards and the cost to attend Northwestern University. Deklan Novak had a spot on the 112-man roster before he arrived at camp.

Novak headed to the main auditorium where the entire team was meeting at summer camp for the first time. There was a pre-camp meeting on the Evanston campus during the week prior to summer training camp. Deklan Novak was unable to attend the team meeting in Evanston when his roster spot was introduced to the team. Novak was training 88 candidates during their boot camp so that hopefully those candidates received an invitation to attend BUD/S training in Coronado, California following boot camp graduation. Novak and one other Navy SEAL were under the command of Master Chief Evan Kerr and assigned to Great Lakes for the purpose of teaching BUD/S candidates how to train for the toughest training in the military. (PO1) Deklan Novak's job at Great Lakes Naval Recruit Training Command (RTC Great Lakes) was to skim the top off the candidate pool and eliminate the Navy SEAL hopefuls that had zero chance to make it through BUD/S, since once accepted into BUD/S training, the average dropout rate exceeded 80%. The NU football team was also told at their pre-summer camp meeting that an active-duty Navy SEAL was a full-time student at Northwestern University. The same Navy SEAL was also a BUD/S instructor at Great Lakes Naval Recruit

Training Command in Waukegan and would be a first-year linebacker for the Wildcats in 2013. The football team meeting room fell as silent as a mid-term final exam at the McCormick School of Engineering.

NU Head Coach Lance Adams played Mike linebacker also known as middle linebacker for the Wildcats from 1993 to 1996. Northwestern was 10 and 1 during the 1995 season and Adams led the school to the 1996 Rose Bowl. Adams was drafted in the fourth round of the 1997 NFL Draft by the Pittsburgh Steelers, but injuries ended his playing career early. NU Head Coach Lance Adams never imagined that Novak would play one down during a game, but once Deklan Novak secured his enrollment at Northwestern, Adams knew the intangibles brought to the locker room by an active-duty Navy SEAL could be invaluable. From one meeting, Adams knew that Deklan Novak carried with him the credo of self-accountability, a concept Adams held as essential to facing adversity. In addition, Lance Adams was fascinated by the toughness necessary to become a Navy SEAL and selfishly, Adams sought out the friendship with Deklan Novak for himself as much as for the team. Lance Adams had three sons, and the lessons Novak could pass down had to make any man a better father, coach, and person. Football skills aside, Adams knew that Deklan Novak had earned much more than a roster spot on his football team. Relentless was a term often associated with a great football player. On the day Deklan Novak asked if he could play football for Northwestern University, Lance Adams had been introduced to what relentless truly was. Awarding an active-duty Navy SEAL a roster spot was not a gift. Lance Adams was honored to have Deklan Novak on the team.

Lance Adams brought Deklan Novak and a former Navy SEAL, David Talbert, previously stationed at (NAVSTA) Great Lakes over to the NU practice facility in 2012 to address the team on what a training commitment meant, looked like and smelled like. Talbert wrote a bestselling biography after he left the SEAL Teams that detailed a troubled childhood, overwhelming poverty, alcohol, and drug abuse that eventually turned into a ten-year stint with the most elite military unit in the world. Navy SEALs went through two years of hell for a lifetime of respect. Adams met with Deklan Novak in his office after Talbert's speech to the team. David Talbert was signing copies of his book for the team. At that meeting, Deklan Novak asked Head Coach Lance Adams if he would consider allowing an older dude to try out for the team next year as a walk-on. Novak explained that he did not have an impressive high school resume and showed the decorated Big Ten Head Coach his credentials.

"Hell yes!" Coach Adams blurted out. Adams was taller than Deklan Novak and twelve years older, but still looked like he could drive a blocking sled through a steel door by himself. Adams was molded in the image of a lifelong Marine although he had not served in the military. The close-cropped hair and square jaw mirrored the image of Tom Highway from the movie, Heartbreak Ridge. "I'll hook you up with the recruiting department on the admissions process." Adams announced. "I am relatively certain that you will have to score a 1300 on the SAT to have a shot at getting into Northwestern. If you get into Northwestern, you will be a part of our team. Walk-ons are treated the same as any scholarship player except their tuition is not part of the football experience. How big are you, Deklan?"

"I am 6' 3" tall and 235 pounds, sir." Deklan replied.

"When did you last play football?" Adams continued. Adams knew he was staring at a linebacker and a special team's player.

"In high school at Loyola Academy in Wilmette. My last year was 1999." Novak thought that might be the deal killer.

"Train hard between now and next summer. Football training is specific. Keep me posted on the training and I'll get updates on the admissions process. I look forward to seeing you in pads." Adams directed, looking at Novak directly in the eyes. Novak knew the Big Ten Head Coach meant what he said.

"Hooyah, sir!" replied Novak and they shook hands. Novak departed that 2012 meeting to find his author buddy, ready to get to work.

CHAPTER ONE

Novak knew his first task was to set out to meet the standards necessary to be accepted into Northwestern University. The consensus was that Novak needed to score between 1300-1500 on the SAT, graduate in the top 10% of his high school class (that ship had sailed), earn a grade point average near or at 4.0 in high school (that ship never arrived to sail away) and submit letters of recommendation from a wide variety of prominent individuals. Of all the admissions requirements expected of him, Novak knew the letters of recommendation would be the easiest to bring to fruition.

Novak was training potential BUD/S candidates at NAVSTA Great Lakes every morning in 2012. Novak's pre-BUD/S classes normally numbered 40 to 50 Seamen Recruits. At the start of each two-week class, Novak asked how many recruits had graduated college. Many hands were raised. Novak then asked how many college graduates graduated from an elite school like Harvard, Stanford, Yale, or MIT. Only a few hands remained up. Novak picked one recruit during each class to tutor him after the boot camp day was completed. There were no perks for the recruit chosen except that a Navy SEAL acknowledged his/her existence.

Deklan Novak attacked his SAT challenge like he attacked any challenge in his life. Novak eventually scored over 1300 on the SAT and was accepted as a full-time student into Northwestern University. Novak was not naïve to think that the letters of recommendation received by Northwestern University on his behalf from two U.S. Senators, and several ranking members within the U.S. Special Operations Command (USSOCOM) did not carry substantial weight. Admiral William McRaven, Commander of (USSOCOM) based at MacDill Air Force Base in Florida possibly tipped the scales for Novak's admission. The Commander at USSOCOM had the following units under command: United States Army Special Operations Command (USASOC), United States Naval Special Warfare Command (NSW) Navy SEALs, United States Marine Corps Forces Special Operations Command (MARSOC), Air Force Special Operations Command (AFSOC), and Joint Special Operations Command (JSOC) to protect and advance United States policies and objectives.

In addition, the resume from Admiral McRaven included the co-command of Operation Neptune Spear with the CIA and the SEAL Team Six op on May 2, 2011, that successfully located and killed Osama Bin laden. Most applicants to college were able to list high school participation in band, drama club, debate club, letter sports and intramural athletics. Novak's application instead had multiple wartime deployments to fight terrorism in Iraq and Afghanistan over a decade, a leadership position at the only Navy boot camp in the United States, and graduation from BUD/S, what most often had been described as the toughest training in the world. The recommendation letters sealed the deal.

Novak's football experience heading into his first season with NU most likely was destined to produce a zero chance to see the field as an active player, while adding to the grueling schedule of a student athlete and an active-duty Navy SEAL. Novak's marriage that eventually produced four boys was disintegrating daily under his blind eyes. Deklan Novak's wife, Amber Bale Novak, moved to San Diego while he was a BUD/S candidate in 2006. After the move to San Diego, Amber and Deklan were married and shared an off-base apartment during the 16-month Navy SEAL training regimen that included BUD/S Phase One-physical conditioning (7 weeks), Phase Two-combat diving (7 weeks) and Phase Three-land warfare (7 weeks) training. Survival, Evasion, Resistance and Escape (SERE) School and SEAL Qualification Training (SQT) followed to complete arguably the most intense and grueling training in the world. The marriage had been rocky during BUD/S and continued to be a roller-coaster during the decade-plus years that Deklan Novak was deployed around the world as a member of SEAL Team One. There were obvious reasons why there were few members of SEAL Team Six, officially known as the Naval Special Warfare Development Group (DEVGRU) Green Team who were married. Finally, a stable assignment to (NAVSTA Great Lakes) came to fruition for Deklan Novak and retirement from the Navy was in sight. Amber had dreamed of the days when Deklan was no longer a Navy SEAL, but she had not planned for the additional time apart that Division I college football required.

CHAPTER ONE

The auditorium at UWP was filled with the entire team, coaches, the Athletic Director Jim Tunney, the Assistant Athletic Director Manny Griffith, nutrition staff, recruiting staff, support staff and conditioning people. Defensive coordinator Behr Thomas was speaking to the team. Head Coach Lance Adams was seated directly behind the podium where Coach Thomas was speaking. The seating aisles rose gradually from the front of the room and the entrance door was off to the side of the main stage. There was a video screen lit up behind Coach Thomas illuminating the iconic stone and iron archway synonymous with the main Northwestern campus and located on Sheridan Road at the apex of the lakefront acreage. Deklan Novak entered the large, tiered meeting space and Coach Thomas stopped talking to see who was interrupting his time before the team.

(PO1) Deklan Novak wore Navy shorts, flip-flops, and a UDT/SEAL instructor tee shirt. Novak had full-sleeve tattoos on both arms, a chest tattoo was barely visible inside the neck of his tee shirt and the new team member had a three-day stubble on a premature gray beard. Novak's arms looked like vein roadmaps on a body builder. Novak's tattoo's included a John Henry tattoo from a Johnny Cash ballad, a tattoo honoring Deklan's brother Danny or DJ who passed away in 2008, a tattoo representing a chaotic world descending into order into a labyrinth into the mind, soul, and self, a bone frog tattoo that was the coveted rite of passage as a Navy SEAL who had done his first deployment. The room was dead silent and the stereotypical image that most young men had of a Navy SEAL, came to life before them like a video game mythical figure larger than Kratos.

The tattooed warrior also wore an image of Virgil crossing the river Styx on his way to pick up Dante, a tattoo of a skeleton angel whispering into the ear of a naked woman, the names of his sons tattooed on his chest, the three-headed hound of hell, and a large tattoo of the four horsemen of the apocalypse. Novak slid his flip-flops across the tile floor to minimize the slapping as he looked for a seat. The team remained silent but took in the new member like they were watching Marvin Hagler ascend to the ring.

"Welcome, Mr. Novak." Coach Thomas announced and did not reprimand the late arrival. Adams had told his staff what to expect and when to expect it regarding Deklan Novak. "We had our little introduction session on campus last week before our annual sojourn to Kenosha. Each player stood up and introduced themselves. Please Mr. Novak, introduce yourself to the team and tell us a little about where you are from and why you decided to attend Northwestern University." Behr Thomas smiled with a snarky and sarcastic grin.

Thomas was less than one year older than Deklan Novak. Barely 33 years old, Behr Thomas had played free safety for the University of Michigan, gaining a reputation for hitting like a freight train. Thomas was drafted into the NFL by the Chicago Bears and like Brian Urlacher, Behr Thomas was moved from safety to outside linebacker as a pro. Behr Thomas put on an additional twenty pounds and thrived on the outside in the Bears 3-4 defense. Seven concussions in five years, however, ended Behr's professional career and led him to the sidelines and to the podium in the UWP. The team waited in silence as Deklan Novak found a seat, turned, and addressed the team.

"Greetings." Novak began. This introduction was going to be short and vague. "My name is Deklan Novak. I grew up in the south suburbs of Chicago and I moved to the North Shore when I was in middle school. I went to Loyola Academy during high school and followed the Wildcats closely as they played fifteen minutes from my house. I joined the Navy in 2006 and enrolled at Northwestern last spring. I got a late start, but I promise not to bring my walker to the practice fields." Novak sat down amid the laughter. When the noise faded, one of the linebackers assigned to room with Novak stood up and asked a question on everyone's mind.

"Are you a Navy SEAL, dude?" Marty Reyes asked and knew the answer. Reyes was from Anderson High School in Austin, Texas and was raised among Texas royalty. His father was a PRCA World Champion Bull Rider, and his great-grandfather was a renowned bootlegger around the Texas capital during Prohibition and made a fortune from purchasing land in Austin after the Great Depression. Football was king in Texas and Reyes was revered in his home state. Reyes clearly wanted to mark his territory early on.

"Yes, I am a Navy SEAL." Novak replied and the next question was as expected as the questions received by Santa with any child on his lap.

"How many guys have you killed?" Reyes followed the script as expected and one that Deklan Novak heard often. Whenever people met a Navy SEAL, the question inevitably arose. Marty Reyes started nodding like a bobblehead doll, turned to his teammates, and waited for his raucous adulation. A few smattering retorts followed in support of the Reyes inquiry. All eyes returned to Deklan Novak.

"None," Deklan Novak stood while he replied and sat back down.

"Bullshit!" Reyes shot back. "Don't be shy my man or what does your shirt say…UDT SEAL Instructor? Power Five football is intense. We are big boys, here. We know you were deployed to Afghanistan and Iraq on multiple occasions. Coach Adams told us you were a breacher on SEAL Team One. You were the first one into the shit on every mission. How many dudes did you blow away? Seriously, dude…more or less than a hundred? Just tell us that much."

"My apologies," Deklan Novak stood back up. "My mind gets a little foggy sometimes. I took part in Combined Joint Special Operations Task Force-Afghanistan, also designated as Task Force Dagger, which was asked to conduct Unconventional Warfare in Northern Afghanistan. We were very good at killing the bad guys. The most difficult task during the war on terror was not killing the bad guys but finding the bad guys. To find the bad guys, we had to trust or depend on intel from the United Nations Security Council Resolution 1386 authorizing the International Security Assistance Force (or ISAF) which was primarily made up from Afghan and Iraqi soldiers working for the United States Special Operations Command, better known as (USSOCOM)."

Novak paused, allowing the acronym vocabulary of the military world to sink in. "The local population did not welcome the United States military. We had built-in enemies within our own command. 9/11 gave Cheney the public appetite to manufacture a war. There were no weapons of mass destruction or WMD in Iraq. We entered Iraq in 2003 to destroy the WMD and kill Saddam Hussein. By the end of 2003, the WMD searches were aborted, and Saddam was captured."

"We remained in Iraq until 2011, building 505 bases occupied by 170,000 men and women in uniform and another 135,000 private military contractors working in Iraq. Money in war is not made by invasion, victory and exit. Money is made by occupation. I'm not sure what your name is my friend, but we didn't count kills, we counted days alive. During Operation Enduring Freedom, we geared up at Forward Operating Base (FOB) Chapman or at FOB Salerno to launch direct action combat missions centered on locating and neutralizing high value Taliban targets." Novak paused again before continuing. "FOB Chapman was named after Sergeant First Class Nathan Chapman, the first U.S. soldier killed by enemy fire during the Afghanistan war in 2002. Improvised explosive devices (IED), caused almost 70% of all coalition casualties in Afghanistan and Iraq. We had to constantly be aware of roadside IEDs, house-borne IEDs (HBIED), and vehicle-borne IEDs (VBIED). Insurgents worked hard at remote detonations and roadside death traps to kill American and coalition forces," Novak paused again, his mind recalling those memories.

"Often, in Afghanistan," he continued, "when we were moving between FOB Chapman and FOB Salerno, we would drive in a two or three truck convoy in the dead of night, fully blacked out with me and the other drivers on night vision goggles at hyper fast speed going nut to butt so we couldn't give the IEDs enough time to detonate accurately. The trucks we drove were Toyota Hilux pick-up trucks with a ton of added armor and 3,500 pounds of hidden equipment on board. Each truck cost the U.S. government north of $300,000 and the vehicles could blend in with local traffic if necessary. The trucks were armored like an M1 Abrams tank, carrying comms and satellite systems designed to shut down all cell phone use in the area and to call down air support from the U.S. led Combined Air Power Transition Force. Our missions usually had an overhead convoy of Boeing F/A-18 Super Hornet fighter jets flying above. The Super Hornets

CHAPTER ONE

carried a standard load of a 20mm M61A2 Vulcan Gatling gun with 578 20 mm projectiles carried for short range work. The eleven hardpoints allowed for a mixed ordinance carrying capability. A hardpoint is an attachment to the mounting frame on the aircraft that carried missiles, bombs, rocket pods and jettisonable fuel tanks. The wingtip launches were typically reserved for 2 x AIM-9 Sidewinder short-range missiles." Novak paused, knowing he was throwing a lot of military jargon and specs at his audience.

"The bad guys couldn't detonate their roadside bombs," he continued explaining, "because we shut down their comms from the trucks. We drove those mother fuckers at 70 to 80 mph on one-lane dirt roads that were no wider than a standard twin bed, in pitch black darkness and we were bumper to bumper in a convoy that never knew if we missed the one comms connection that might blow us into a pink mist. We were big boys too, my friend, so I guess we both can relate to intense." Deklan Novak sat down for a third time. The auditorium was as silent as a country road at four o'clock in the morning.

"Drop it, Reyes." Dusty Lindell stood up and addressed their starting Mike linebacker. Dusty Lindell was the Junior starting fullback and one of the team Captains. "When Deklan Novak is ready to talk to you or to anyone else on this team about his experience in combat, then he will do so on his own terms. Welcome, dude." Lindell addressed Novak directly. "Don't mind Marty, he wears a Call of Duty tee shirt under his uniform on game day."

"Good call, Dusty." Behr Thomas took back the reins. Deklan Novak did not respond in any way. "Shut the fuck up, Reyes, and learn what you can from Mr. Novak. I am certain he has been through training that we can only imagine. Hell Marty, Mr. Novak may bring you a Navy shot glass from the commissary at Great Lakes if you're nice to him." Coach Thomas smiled and continued. "Thank you, Deklan. We all appreciate your service. Our sole goal here is to prepare and win the Big Ten title. Anything that occurs after that must be precipitated by a Big Ten title. If anyone in this room doesn't plan to win the Big Ten title this year, please pack up your shit and go home. I'm not here to monitor a juvenile territorial testosterone contest. I know Coach Adams doesn't give a shit if you, Marty, want to wear your Lone Star crap on your sleeve, but here we earn our positions and our playing time. The starting lineup is not decided on by a high school state title or by how many recruiting letters you received since your acne started to recede. Positions are earned on the field and that starts at 6 a.m. tomorrow morning. My defense is a meritocracy, boys. Reyes, you can ask Mr. Novak what that means." And with that, Coach Thomas gave the floor back to the head coach. Marty Reyes did not know what to think of his new roommate and he had no idea what a fucking meritocracy was.

"I hope all of you get the chance to play in the NFL." Head Coach Lance Adams took the podium and began his closing remarks. "The reality is that most of you will not play in the NFL. My job may appear to be molding the best football players I can mold over the next four years or however many years you have remaining on our football team. That is not my job. My job is to mold the best men I can mold over the time you spend on this football team. This is not about the next four years. It's about the next forty years." Coach Adams paused. The coach looked over his team as everyone remained quiet. He folded his hands and brought them to his mouth and continued. *"Aristotle claimed that character develops over time as one acquires habits from parents and community, but first through reward and punishment. The conclusion is simple. Character is learned and the harder you fight for it, the better men we become.* I agree. Character is the only thing not driven by money. Genetics have nothing to do with it. You cannot steal it, find it, or fake it. Thank you for the privilege to coach you all. Now, get ready for tomorrow because you will be taught to get comfortable being uncomfortable and you will know where that came from soon enough."

Lance Adams stepped away from the podium. Behr Thomas reclaimed the podium and waited for his head coach to exit the auditorium.

"When will we know if we are good enough to win?" Thomas began and paused as he side-eyed Deklan, as if to say that Novak wasn't the only BMF in the room. "Walter Payton was quoted as saying…when you are good at something, you'll tell everyone. When you are great at something, they'll tell you." Coach Thomas spit on the floor. He looked around and then took a large mouthful of water from his water bottle as the players watched. Thomas swirled his intake, and then spit out a torrent of saliva-infused liquid onto the stage, threw his water bottle, half-full, into the players. He had one more message and he screamed at his team. "We are not going to tell anyone this year about how good we are. Starting tomorrow, all your communication will be done with your pads, your helmet and your fucking laser focused resolution to kick seven shades of shit out of the team or the man in front of you!" And with that, Behr Thomas followed Coach Adams out of the lecture hall auditorium.

The UWP auditorium exploded with a deafening ovation. The entire entourage of large, agitated Big Ten football players began jumping out of the seats they occupied. Everyone left in the room unleashed a plastic water bottle tsunami against the still illuminated front wall video screen, while chairs were slammed down like car compactors, and the inspired throng began a primeval yelping call depicting a territorial coyote that was marking his province in the Sonoran Desert. Northwestern summer camp had officially begun.

CHAPTER TWO

2013 STEAMBOAT SPRINGS, COLORADO

THE YAMPA VALLEY MEDICAL CENTER was a 34 bed acute care hospital that provided sophisticated medical services to more than 5,100 outpatients annually. Raya Nolan Thomas sat in the emergency room waiting area after bringing Kaley to the ER at 7:00 a.m. Kaley had been vomiting most of the night and was dehydrated, not able to keep anything down for the past three days. This was their second trip to the Yampa Valley Medical Center ER in the last eleven months. In between the two visits to Yampa Valley Medical Center, there were visits to Vail Health Hospital (93 miles from Steamboat Springs), Edwards Medical Campus (80 miles from Steamboat Springs), Dillon Health Center (91 miles from Steamboat Springs), and High County Healthcare in Silverthorn (89 miles from Steamboat Springs).

Raya and Kaley were waiting for a triage assessment to be conducted. Raya knew the nurses would ask if Kaley had been to the ER before. The prior visit to The Yampa Valley Medical Center ER was inconclusive as to the cause of Kaley's symptoms, which were identical to her current condition. All the tests run during the first visit came back as normal. The emergency room physicians had a sick child admitted for observation, to stabilize her ability to eat, to administer IV fluids and all the tests run during Kaley's first stay at Yampa Valley MC, failed to explain why a sick child tested out as completely normal. After two nights, Kaley improved, ate normally, and was sent home with an undetermined diagnosis as to what caused her to get sick. The triage nurse would review the past visit, but the staff would have no way of knowing about the ER visits to Vail, Edwards, Dillon and Silverthorn, and any other visits to acute-care medical facilities within a two-hour drive time from Steamboat Springs. Raya Nolan Thomas brought Kaley back to Steamboat Springs after the divorce was final. The transient resume of an assistant coach in college football secured sole custody of Kaley to Raya. Behr Thomas had unsupervised visitation rights with sleepovers and vacations, but Behr had no legal or physical custody rights. Raya grew up in Steamboat Springs, Colorado.

Steamboat Springs, Colorado is in Routt County located in northwestern Colorado. The winter ski resort community had a recorded population year-round in 2010 of 12,088 at an elevation of 6,732 feet above sea level. Known as "The Boat", Steamboat Springs is 150 miles northwest of Denver and has beautiful summers contrasted with brutally long and cold winters. The city is almost 90% white and the median price for a home in Steamboat Springs exceeded $500,000 in 2010. Raya Nolan Thomas and her live-in boyfriend Carmen Gallardo were not part of the affluent residents in the Boat. Raya's mother had recently passed away and her father was long gone. The parents divorced when Raya was a child. Raya managed to attend Colorado State University in Fort Collins, Colorado, some 175 miles east of Steamboat Springs. Raya took out Federal Student Aid loans and found in-state scholarship help to attend CSU. There, she met Behr Thomas, who played football at CSU briefly before transferring to the University of Michigan.

Ray's mother left no money for her only child, but Nancy Nolan left Raya the family home, a small three-bedroom trailer in the Dream Island Mobile Home Park on the Yampa River in the heart of downtown Steamboat Springs. The main internationally known ski area in

CHAPTER TWO

Steamboat Springs is on Werner Mountain about six miles from downtown. There is a much smaller ski area based downtown on the Yampa River called Howelsen Hill Ski Area. Howelsen Hill Ski Area, opened in 1915, is Colorado's oldest ski area. The Dream Island Mobile Home Park was only blocks from the main downtown area in Steamboat, where all the trendy restaurants, ski-shops, bars, and coffee joints were located. Dream Island Mobile Home Park sat a few hundred yards from a natural sulfur hot springs pond that the locals called Stinky Pond. The palatial estate needed some work, but there was no mortgage and the real estate taxes were the only expense of consequence to living there.

Raya worked as a part-time server at 8th Street Steakhouse in downtown Steamboat Springs. Carmen had a part-time job at the Howelsen Rodeo Grounds and the Brent Romick Arena. The venue hosted a summer professional rodeo series each Friday and Saturday night. Carmen ran the Bobcat tractors that cleaned and groomed the grounds. During the nine months when the rodeo was not operating, Carmen caught some work at the venue for different events and maintenance, but Carmen spent most of his time during the long winters in Steamboat at the local taverns and sold cocaine to the tourists. Cocaine in an affluent tourist destination like Steamboat Springs was a cash machine on steroids. A kilogram of cocaine from Columbia was purchased in Columbia for $2,000. By the time the same kilo hit the United States border, the value had jumped to $30,000. When the kilo was broken down into grams, and sold on the street, the value jumped to $100,000. That is a 4,900% increase. Gallardo did not factor into the equation until the final passage. Gallardo was only a street dealer with a lengthy record. Gallardo and Raya Nolan met each other at Steamboat Springs High School, Class of 2001. Raya had been in contact with Carmen Gallardo during her divorce.

Raya Nolan Thomas retained health insurance for Kaley from her ex-husband and his position as the defensive coordinator for Northwestern University. Raya and Kaley were finally led to the triage room. The ER staff started an IV line to get some fluids into Kaley. The questions had all been heard before.

The triage nurse decided if the patient was sick and how sick. In triage, the nurse will assign the patient an acuity score based on the Canadian Triage and Acuity Scale (CTAS), where 1 is the most urgent (cardiac arrest, major trauma) and 5 is non-urgent (diarrhea or a bandage change). Based on the triage evaluation, the nurse decides where to send the patient for treatment in the ER, the trauma room, an acute treatment area or to wait for out-patient instructions before going home.

"When did you first notice that there was something wrong, Mrs. Thomas?" The nurse asked and Raya did not correct the name or clarify.

"Kaley hasn't been able to keep anything down since Monday." Raya lied. "I had hoped this was not another incident like last year. Maybe she just had a flu bug or something."

"Did Kaley ingest anything unusual that you know of?" The nurse asked. "Did you take the family out to dinner anywhere in the last day or two? Did you eat at someone's house?"

"No. Kaley has been at home all week." Raya replied. The questions were then directed to Kaley. The nurse was trained to not only listen to what the child had to say, but to observe how she answered each question. Was the child scared? Was the child rehearsed? Kaley was scared, as expected, but none of the answers seemed to be rehearsed or forced. There were no visible signs of bruising or injury. If the triage nurse suspected any form of child abuse, the nurse was obligated by state law to separate the child and the parent or parents to ask the child questions away from the parent. There were no red flags in Kaley's case that were evident to the triage nurse. Raya Nolan Thomas appeared distraught and worried about her daughter, nothing unusual or suspicious about concern. Parents that cannot explain broken bones or physical trauma to the child are immediately subjected to the police and the Department of Child Safety in that state. Parents unable to explain an illness are often matched with doctors who cannot explain the illness.

CHAPTER TWO

"I need to examine your daughter now." The nurse explained. "We will establish her vital signs, run blood tests and then run some tests in order to see what may be causing the vomiting and inability to eat."

"Of course." Raya uttered barely audible. "I just want to see her feel better." There were tears in Raya's eyes. The triage nurse came over to Raya and assured the worried mother that Kaley was in good hands now.

Raya complied quietly and did not protest anything. Raya Nolan Thomas, 30 years-old, was a pretty woman. At 5'5" tall and 110 pounds, she had sandy blond hair that hung disheveled on her shoulders but looked attractive in a non-intentional seductive manner. Raya was naturally chic thin but with a granola, trendy air of health, a fashionable, desirable figure that men called a gapper and most women envied. The last year had brought on a noticeable change to Raya's appearance. There was a visible gaunt, emaciated nuance to Raya's otherwise striking look.

Raya never understood the problems in her marriage to Behr Thomas. Behr had married out of his league. Behr Thomas, as many had so poetically illuminated in a football vernacular how ordinary men end up with pretty women, had outkicked his coverage. Raya wanted to marry a successful man and Behr had told her many times that he was going to be a head coach in a Power Five conference before he was forty. Power Five head coaches became wealthy. Behr didn't tell his new wife that he would rarely come home, the daily hours for an ambitious assistant coach left virtually no time for the spouse or family, and the clock on a college football season was year-round. The hours were horrible, and the money wasn't great, either. Assistant coaches in college football moved more often than military families. Raya Thomas hated living in Evanston, Illinois. Raya hated Ames, Iowa when Behr was the secondary coach for the University of Iowa, hated Charleston, Illinois when Behr was a linebackers coach for Eastern Illinois University. The couple argued almost from day one. It was late October. The weather had begun to turn colder in Steamboat, and Raya wore a North Face pink down vest over another lighter nylon white jacket, a pair of Wrangler jeans and cowboy boots.

Raya looked like a "Boat" native. Raya drove her daughter to the emergency room. She knew the way.

Kaley Thomas had just turned six years old. Kaley was 42" tall and weighed 25 pounds, almost ten pounds or nearly 30% below the lowest weight listed for a child her age. Kaley looked like an angel from Latin America, Azerbaijan, or Armenia. Kaley had light-brown skin, long straight dark hair cut to the same length, parted in the middle, and looked like she had green, emerald gemstones implanted in her eyes. One side of her hair was tied back with a white barrette. Kaley's face was stunning. Nurses and doctors in the hospital instinctively looked once as they passed and took a second look to make sure. The second looks were bilateral in that trained medical professionals couldn't define the young, undernourished patient. She looked scared but calm, in pain but not crying. Kaley wore one-piece pajamas with feet that were stained by the illness that had taken residence within her body.

The tests took longer than most of the other visits, but Raya never mentioned that to the staff at Yampa Valley Medical Center. Tests included blood work, urinalysis, abominable x-rays, a CT scan, MRIs of the head and lumbar regions, an upper gastrointestinal endoscopy (GI) and an ultrasound. Raya kept pacing while Kaley was out of the room for tests. There were more than a few terse, distraught mother visits to the nurses' station to ask when Kaley was coming back. Kaley was eventually brought back to the treatment room where they had been assigned.

The emergency room physician oversaw the ER. The position of Emergency Room physician had evolved over the past two decades into a very sought-after specialty. As an emergency room physician with specialized 3-4 years of residency done solely as an emergency room doctor, the job offered regular shift work, limited on-call requirements, and no follow-up with patients. Emergency room patients were referred to other doctors for follow-up care and diagnostic testing. The flip side was intense pressure during trauma and mass trauma. The ER doctor came in to talk to Raya.

"We are going to keep Kaley for at least a couple days." Dr. Colton

Wade told Raya Thomas. "We have to know that she is keeping food down and we want to see her get some nutrition over a few days." Dr. Wade spoke slowly. Dr. Colton Wade was 44 years old, average height and build, but had a kind face with a great head of prematurely white hair that was longer than most imagined appropriate for a doctor. Dr. Colton Wade resembled a thin, younger Jeff Bridges.

"I've been trying to get her to eat, but she can't hold anything down and now, she doesn't even want to try." Raya told the doctor.

"That is not a surprise. That's why we will keep her here for a few days." Dr Wade continued. "I believe your daughter has gastroparesis. That's a condition that prevents the muscles in the stomach from moving the food through the digestive tract. The stomach can basically shut down. We see this produce episodic fits of vomiting where the patient eventually becomes afraid to eat. The patients are often nauseous and sick to their stomachs. We don't know exactly what causes gastroparesis, but it can be caused by nerve damage, something blocking the digestive tract or diabetes. Your daughter does not have diabetes. Let's get Kaley comfortable and get some added nutrition into her system. She will feel better quickly. I'll come by tomorrow and I will have had more time to evaluate all the tests, and I should have a clearer idea of what has caused this and what we will do to fix it." Dr. Wade turned to Kaley. "Does that sound okay with you, Kaley?"

Kaley nodded. That was going to be okay with Kaley. Kaley was thinking about train rides and chocolate ice cream sundaes on board inside a dining car. There was a breathtaking train ride between Durango and Silverthorn that was originally built in 1882. The Durango & Silverthorn Narrow Gauge Railroad (DSNGRR) traveled through the San Juan National Forest as it followed the stunning Cascade Canyon. Kaley and Raya had watched the vintage train along the mountains above Silverthorn during a visit to CHPG High Country Healthcare in Silverthorn. Raya told Kaley that they would take that train someday together and have a special lunch on board. Her mother told Kaley that they would see big horn sheep on the mountainsides as well as majestic bald eagles soaring high in the sky like untouchable angels for the wilderness. Raya lied about most things to Kaley.

"How is your health holding up?" Dr. Wade asked Raya Thomas and startled her.

"My health is not the issue here, but thanks for asking." Raya replied. "I am worried about my daughter but otherwise fine."

"You told the nurse in triage that you are divorced." Dr. Wade paused briefly but his comment was not a question, and he was not waiting for a response. "Do you have legal custody of Kaley? I am sorry to ask that, but we are required to establish who has legal custody so we can proceed with medical decisions without court approval. Is Kaley's father aware of her medical issues and do we need the father to be legally involved in Kaley's care?"

"I told the nurse that I have sole custody of Kaley, and her father is not involved." Raya answered, more than mildly concerned that the subject had already been asked and answered.

"Of course." Dr. Wade followed up. "Do you have a medical background? Nursing school or military service as a medic?"

"No. I have no medical background." Raya answered.

"Your care for Kaley has been spot-on. That's the only reason I asked." Dr Wade explained. Raya knew why the question came up. The kind-looking emergency room doctor was wandering off-course and reaching. Raya Nolan Thomas curled up on the reclining chair next to Kaley's bed.

"Try to get some rest tonight, Mrs. Thomas." Dr. Wade spoke as he was leaving the private patient room where Kaley and Raya were about to spend the night.

CHAPTER THREE

2000...HIGH SCHOOL GRADUATION, CALIFORNIA, THE MMA, AND THE MILITARY

GRADUATION FROM LOYOLA ACADEMY in Wilmette, Illinois was a dead lead-in to college for most of Loyola's graduating seniors. The tagline for the school's description was a private Jesuit college preparatory school. The summer following a senior year at Loyola was a summer spent getting drunk with friends who were not going to be around much moving forward. Everyone pledged in the yearbooks that the friendships were to last through eternity, but reality rarely cooperated. The year was 2000 and Deklan Novak had no plans to go to college. Nathan Garfield, Deklan's best friend at Loyola and a fellow football player, and Deklan had teamed up during the summer following their junior year at Loyola and started a small patio construction business. Nathan was very knowledgeable in the industry and had learned the basics from his father's business. Nathan's father was a territory construction foreman for a major home builder in Cook County called Surrey Hills Homes. Deklan and Nathan poured simple patios and built fundamental back yard decks while slipping the projects under the radar without permits. Nathan's dad helped to provide job leads for the boys.

The work was always done behind the homes and their trucks were marked with the name, North Shore Back Yard Contractors, and a phone number. Most neighbors and city maintenance crews assumed the business was one of the countless landscapers that blanketed the northern suburbs. These were the communities boasting 30 MPH speed limits on major thoroughfares and traffic jams clogged each morning not with commuters but with landscaping trucks and their trailers bursting with Mexican lawn caretakers. Union cards were as scarce among the landscapers on the North Shore as locating a sherpa in Kansas. Nathan drove a 1990 Ford F-150 and Deklan drove a 1985 Ford Bronco, full sized SUV, with a 351 cubic inch, 5.1 L V-8, 3-speed automatic on the column that he purchased for $2800 in 1999, mostly from the money made from the patio business. The three-door utility vehicle's odometer read north of 155,000 miles, but the beast ran like a tank and got the job done.

After graduation, the boys put their heart and soul into the patio business. Nathan was looking for a long-term career. Deklan was restless and decided after three months that his calling was California, acting, fame and fortune. Deklan Novak had last acted in middle school but recalled how good he was. There was an audition setup by Deklan's mother in 1992 for the lead role in Free Willy, a 1993 film by Warner Brothers that grossed a remarkable $153.71 million…a bonafide box office hit. Deklan was ten years old. The Chicago Film Institute located at the School of the Art Institute of Chicago (SAIC), regularly held auditions for major film projects looking for new young talent. Deklan auditioned for the lead role of Jesse and had three callbacks but ultimately did not get the job. There was no acting for Deklan in high school because athletics ran his life, but the decision was made to take California by storm with zero plans in place, no connections, a beat-up Ford Bronco, and very little money. Deklan bid his buddy farewell and split for San Diego at summer's end. There were a few friends in San Diego and that seemed close enough to Hollywood for a start.

Life in San Diego for Deklan Novak became a vagabond existence from couch to couch and Deklan was a happy camper. Periodic trips

CHAPTER THREE 27

to Mexico provided more fun and ate up much of the remaining cash. For some income, Deklan worked during September and October 2000 for a haunted house company that set-up spooky backdrops for Halloween parties and events in Del Mar, Coronado, and Rancho Santa Fe. Deklan also managed to get a job driving a rickshaw, a two-wheeled passenger cart, pulled by a pedicab driver. Tours covered San Diego Embarcadero North, Harbor, Seaport Village, Petco Park, USS Midway, Cruise Ship terminal and the Maritime Museum. The gig was good exercise and there were plenty of tourists to pull around San Diego. The undisciplined trip began to wear thin. Money was tight and after finally moving in with some family friends in San Diego, the future began to look cloudy.

The Bronco was giving out and Deklan sold it to make enough money to head up to Los Angeles and find the agents waiting for him to arrive. Deklan did not have a personal photography portfolio, did not have an acting resume, and did not know anyone in the business. Deklan cold-called a dozen acting agencies and a dozen modeling agencies. The city did not jump in unison with excitement to sign the muscular eighteen-year-old to an exclusive contract for representation. There were zero contracts offered and the instructions on how to break into the movie, television and modeling business in Hollywood began and ended with portfolios and experience. Deklan Novak ended up back in the Chicago area before the holidays in 2000. Nathan welcomed Deklan back to the patio business, although the winter in the Midwest put a serious halt to the patio building business. Novak was nearly broke again, was not famous and needed a new plan. Becoming a professional carpenter was great for the time being but that was not the career Deklan envisioned for himself.

After a few more cold months, Nathan decided that the two partners needed to learn how to build real houses and how to become certified carpenters. Those crews worked year-round. Nathan talked his way into apprentice carpenter jobs for Deklan and himself at a major custom home builder that Nathan's father had done business with inside the Northern Illinois counties of Cook and Lake. By 2001, both boys were working hard days of construction on a major home building crew and learning how to build homes from the ground up. Nathan had found his niche. Deklan remained restless, although he liked the physical labor and the crew of men he worked with each day. The terrorist attacks on September 11, 2001 hit while the boys were working on a home in Deerfield, Illinois. The radio reports had virtually shut down the job site, while the crew huddled to find out more information. Finally, the crew chief made them go back to work through his informative wisdom.

"Get back to work assholes." Crew chief Manny Gomez barked out louder than a megaphone. "I guarantee you that you can't do a God damn thing about what's going on, so back to work boys. Listen to the radio at lunchtime."

As the months went on and Deklan began to see some of his friends go off to war, the image of what he was supposed to do began to form. Deklan Novak watched his goofy friends join the military, either in the National Guard or the Army and head off to war. Some came back as grown men with PTSD and all the goofy was gone. Deklan Novak wanted to know what those men now knew. Deklan believed in the nation as a beacon of freedom, and that Americans had the right to live free from terrorists who were willing to kill themselves to kill many more innocent civilians. Deklan looked into the recruiter's offices near his home. The first thoughts were to join the National Guard like some of his friends, but that seemed to be a noncommittal approach in Deklan's mind. The National Guard was a service commitment with minimal active duty mandatory and regular reserve duty obligations. Wartime circumstances increased the chances for National Guard troops to be called in to active duty but most of that duty was not front-line duty. Many states have challenged the deployment of National Guard troops without a declaration of War by Congress.

The Army Rangers presented a better option for Deklan to find the action and the Green Berets exceeded the Rangers in their scope of responsibilities and their path to the battlefields. Green Berets were the actual U.S. Army Special Operations Forces. Army Rangers were used as a rapid deployment force. Green Berets were more specialized and trained longer. Green Berets were paid more than Army Rangers. Both were elite special operations forces within the military, but Deklan leaned towards the Green Berets. That was, however, until Deklan saw the poster for the Navy SEALs at the Navy recruiter's office. Deklan spoke to the Navy recruiter, who basically tried to tell Deklan that the Navy SEALs were probably not a good idea. The chances of even getting accepted into the program were slim to begin with. To earn a Special Operations (SO) contract, the individual had to prove their physical ability by passing multiple Physical Screening Tests (PST) and applicants had to achieve a required minimum score on the Armed Services Vocational Aptitude Battery (ASVAB). Once accepted and awarded a SO contract, the chances of graduating BUD/S training were very slim. Less than 20% of the men who made it to the Navy Amphibious Base in Coronado, California graduated from the training. Deklan knew he could qualify physically to get a contract to BUD/S, but Coronado training was another animal.

Deklan's family was not supportive of his intention to join the military. Deklan's mother knew her son would end up going to war, and she did everything possible to change Deklan's mind, including the offer to pay his tuition to attend college. That was an offer not previously made. The prodding worked and Deklan decided he would try college and go back to playing football.

Deklan Novak took 2-3 courses at Harper Junior College in Palatine, Illinois to be able to attend North Central College in Naperville, Illinois as a full-time student in the fall of 2002. Deklan tried out for the football team as a walk-on player during the same time he enrolled as a student. The football team was a blast, but Deklan found he had zero interest in going to classes. There was an interest in drinking with his teammates and friends but the academic diversion from the military was a resounding failure. While contemplating one night, the demise of his college tenure, Deklan and Danny were watching a UFC match on television.

Deklan Novak was bored and downright depressed. Deklan had gained 35 pounds since high school and was overweight for the first time in his life. There was no direction or fire in his life. Nothing was exciting. Deklan was living at his mother's house with his younger brother Danny, working construction and spent most evenings watching television with Danny, eating nachos and devising self-delusional ways they could outsmart society and avoid the boundaries associated with their current state of poverty. Binge drinking and gambling with a work hard, party hard attitude was rooted within Deklan's deep-seated anger and resentment from a fatherless upbringing. Every weekend was a serious liberty risk. Fights, driving while annihilated and the general misogynistic contempt for his female encounters showed how Deklan viewed the world around him. The military had faded from his passionate pursuits and had joined the growing list of failed attempts, acting, modeling, athletics, and school. Deklan had accepted defeat upon his return from Southern California. It seemed the time had come to play out the middle American existence by working hard and eventually starting a small family business. Deklan had rolled the dice and failed. Go big or go home. Deklan went home.

One unsuspecting evening, life took a detour. Deklan and Danny were eating nachos, and the cheese was dripping from Deklan's burgeoning second chin while they were watching a UFC Light Heavyweight Title fight between Chuck Liddell and Randy Couture. The fight was THE fight of the year in the UFC! Danny and Deklan lost their minds cheering and screaming at the action. That night, Deklan dreamed again about the glory of battle at that level. What was it like? What was the battle like for the classmates that Deklan had gone to high school with who had joined the military? Deklan had first decided after 9/11 that he needed to experience combat. Watching Chuck Liddell fight Randy Couture brought Deklan back to the precipice within his life. What was it like to step into a cage like those two men did? Deklan needed to know what it took to reach that stage, what it felt like to know victory, what it felt like to know defeat. All Deklan knew now was that he needed to know once and for all, what he had inside of himself. What was the constitution within his heart? Besides war, which was not on the immediate table, the battle inside the cage was real and had few rules. Did he have the mettle to become a part of that world? Deklan Novak postponed his arrival into the Grant Wood American Regionalism Art landscape. American Gothic was about to stand down for the real-life edition of the American Gladiators.

Deklan Novak had decided to become a professional fighter. The diet began the next day. Deklan only ate nachos, cheese, vegetables, and fruit. Laird Hamilton's Surfer Diet and Exercise Plan may have been a healthier and more focused dietary direction, but Deklan's plan was in progress. Deklan drank only water with his unorthodox meal plans and worked out three times a day. Weights once a day and two runs each day. The morning run was a long 5–6-mile run, and the evening runs were sprints. The 35-pound weight gain since high school began to drop off. The goal of becoming an MMA fighter grew into an obsession.

Deklan joined the local fitness gym called Anytime Fitness. Deklan joined the YMCA and began strength training and cardio training at both locations while working the construction job during the day. Deklan sourced the internet to learn everything possible about the MMA and how to become a Mixed Martial Artist. From his days as an elite wrestler in high school, Deklan was able to land a part-time gig as an assistant wrestling coach, unpaid at a local high school. Deklan had to get back on the mats training regularly and the job offered the opportunity. Deklan coached, taught, and trained high school wrestlers while preparing himself for the eventual entrance into the octagon.

It became obvious to Deklan by 2003, that guys like B J Penn, a Brazilian Jui Jitsu MMA Champion and men like him were the future of combat fighting. In the early 1990's, Ronon Gracie created an eight-man single-elimination tournament for the purpose of showcasing the effectiveness of Gracie Jui Jitsu against other martial arts. Ronon's younger brother, Royce, served as a combatant in the tournament and would go on to win more UFC tournaments. Deklan Novak knew what his path in combat sports had become. Deklan dedicated his MMA laser to becoming a highly trained attacker no matter the size and strength of his opponent. It was also apparent that the head coach of the wrestling team Deklan was assisting, Josh Brooks, was a full-time Brazilian Jui Jitsu instructor coaching wrestling as a side job. Deklan joined Josh Brooks at his gym and began to train in BJJ. The training was as enlightening as it was tedious and grueling. Deklan learned the basics of ground fighting and developed the confidence to prepare for the next move. Josh brought two other men into the mix to assist in Deklan's quest to become an MMA fighter. Dean and Kirk Russell owned and ran a small MMA gym in North Chicago called Duneland. Josh called his friends to set up an appointment for the men to look at what Deklan had to offer. The MMA gym in North Chicago rolled from 5-8 p.m. each night. Deklan was set up to stop by the next night at 8 p.m. Deklan was scared to death but excited as hell. He knew he was in great shape by then and was ready to fuck some people up. Sleep did not come easily on the night before Deklan's audition at Duneland before Dean and Kirk Russell.

After a light morning workout and a full day's work, the promise of sheer suffering loomed ahead. Deklan had researched the new gym and the owners. Duneland did not have a presence on Google or the internet. Deklan found the address eventually. The gym was located on the bad side of North Chicago, which was like choosing the bad side of Gary, Indiana. The gym was in a rundown shopping strip mall and the windows were either too dirty to see in or they were covered in condensational sweat. Deklan sat in his car in the parking lot and reviewed his new career decision again. The door swung open and a behemoth of a man, shirtless, resembling the Rock, came outside. The man had a massive tattoo of a black cross that covered his chest and Deklan and the beast locked eyes briefly. A head nod from the Rock followed and Deklan's first encounter with Duneland was history. Deklan entered the gym.

Deklan recalled how the place smelled like a rotting dumpster. The stench filled the air with sweat moisture that made it harder to breathe. Deklan was certain that he wanted to skip getting his mug smashed into the grotesque mats along the floor that had to be the source of the pungency. Walls were covered in dingy mats up to a five-foot level. The Russell brothers built their gym from donations

CHAPTER THREE

and scrap yards. They accepted any exercise equipment and matting that was still marginally intact. Deklan was introduced to the MMA fighter brothers. They asked if he brought his gym clothes.

"Yes, sir." Deklan replied.

"Go get them and throw them on. Let's see what you got." Dean told the new prospect.

Deklan was a polished 195 pounds by then. Deklan had spent months in the weight room, on the treadmill, and in front of the mirror. Deklan Novak looked like a fighter and that was important to him, not his soon-to-be fight instructors. The brothers were not physical specimens or muscle magazine models. Both men looked like blue collar workers buying a Snickers and a Diet Coke at the gas station. Deklan entered the ring with Kirk Russell first. They exchanged a hand slap. Kirk shrugged and twitched his nose slightly. Deklan's upcoming opponent looked completely unconcerned. The popping of focus mitts from the other side of the gym stopped. The eyes in the gym wanted to watch the coming onslaught. Kirk Russell had a professional MMA record of 15-8-2. Many would have called his record mediocre, but Kirk Russell didn't judge his mettle by his MMA record. He did care about being the toughest and most methodical fighter in the room. Deklan moved slowly and under control. Deklan wasn't sure about how live the fight was going to be but decided not to wait for instructions. Deklan moved in quickly to grapple with his instructor. A takedown was thwarted effortlessly by Russell and Deklan found himself standing against the fence fighting from a clinch position with a professional UFC fighter for the first time. Kirk Russell did whatever he wanted to do with Deklan Novak. Novak shot double leg and single leg takedowns only to watch Russell spread out into a perfect counterattack that landed Deklan on the bottom each time. Dean Russell took over for his brother and he worked Deklan into a human ball of disorientation. The demonstrations continued until the brothers finally tired of mopping the floor with Deklan Novak. The torture lasted an hour.

"See you tomorrow at 5 p.m." Kirk said as they left the gym. Deklan was ecstatic and exhausted.

Deklan came back daily for six months of grueling sessions at Duneland. The Russell brothers helped train Deklan in Brazilian Jui Jitsu, boxing, kickboxing, Muay Thai, judo, and Tae Kwon Do. Deklan trained daily and added in daily weight room workouts while also working back on the home building crew. Deklan did not have a manager and was getting restless. He took on the look of his MMA hero, Chuck Liddell. The mohawk haircut caught the family by surprise but was a hit on the construction site. After six months, Deklan inquired about someone scheduling his first fight. The brothers did not think Deklan was ready yet. Deklan did not agree. Deklan had been holding his own at Duneland and if they were not going to manage his new career, he decided to manage himself. There was a well-known MMA tournament venue in Cicero, Illinois. Deklan Novak made a call.

Without a manger, the fighters had to set up their own fights. Deklan called Douglas Penn, a well-known MMA fight promoter in Cicero, Illinois. Penn began his martial arts training in the early 60's with boxing, wrestling, judo, Taekwondo and Jiu-Jitsu. Penn was honored and inducted into the United States Martial Arts Hall of Fame for competing internationally in many disciplines for over four decades. A ninth-degree black belt, Penn fought some of the biggest names in the industry including Carlson Gracie Jr.

Penn asked if Deklan had a manager or a record of note. Deklan did not have a manager, and he was seeking his first fight. After explaining to Douglas Penn where he trained and who his trainers were, Penn agreed to put Deklan on the upcoming card in two weeks in a tournament called Combat-Do. Douglas Penn was promoting an up-and-coming MMA fighter with a 4-0 record and the prospect of another victory via the new rookie named Deklan Novak, was perfect. Penn explained the procedures, the medical requirements, the weight requirements, and who the opponent was going to be, a guy named Tommy Radcliffe. Penn offered nothing more on Tommy Radcliffe, outside of his name and his MMA record. Deklan's first MMA fight was going to be against a polished boxer and an undefeated MMA fighter in the octagon.

CHAPTER THREE

MMA rules called for a 30' Octagon ring with 746 square feet of fighting space. The chain fence around the ring was 6 feet high. Non-title events were 3, five-minute rounds. Title events and UFC sanctioned main event fights were 5, five-minute rounds. Scoring was on a 10 point must system where the round winner received 10 points, and the round loser received between 9 and 7 points. Fouls were assessed for grabbing the fence, holding gloves and clothing, head butting, biting, spitting, hair-pulling, eye-gouging, timidity, and striking an opponent when in the care of a referee.

The Combat-Do tournament was held in a 5,000-seat auditorium. There were 13 matches on the fight card for the night that Deklan had his first MMA fight scheduled. The 26 fighters were split up into two groups that were opposing each other. Deklan had been training for six months. Dean and Kirk Russell were not pleased that Deklan had set up his own fight without their input.

"Do you know anything about your opponent?"

"No, nothing." Deklan admitted.

"Let's look him up." Dean Russell was intrigued by Deklan's initiative. "Fuck it. We'll get you ready."

Deklan Novak told no one else about his first fight. He mentioned it to the crew on the construction site, but knew they had families and were not going to come to Cicero late, after 10:00 p.m. on a Saturday night to see Deklan get his brains knocked out. Deklan wanted to take everything in about his first taste of combat mano y mano as a professional fighter on his own. Deklan didn't want to explain anything to anyone, didn't want to hear stupid questions or concern himself with anyone worried about his safety or experience level.

Deklan was terrified. The fear and apprehension took life from Deklan's chest and the need to remind oneself to take breaths occurred often. Late the night before his first fight, Deklan drove out to Cicero to officially weigh-in. The huge facility was dark and empty, and Deklan only saw a few other fighters and a referee. Deklan tipped the scales at the necessary weight and barely slept that night. On fight night, the arena was a bevy of frenzied activity, filled with intensity. Every light in the building was fired up. The venue was buzzing from the front concession stands to the stadium seats with hundreds of people. Staff, fighters, medical personnel, fans, and media left Deklan with no idea what was going on. Novak walked right in the back door like he belonged. Novak carried his gym bag loosely hung across his shoulder and began to wander. What a delight, he thought. What a terrifying delight. Novak watched a few guys from the top of the auditorium bounce around the metal octagon cage on the stage. They were making sure it was fully assembled and ready for battle. Novak figured he should soak up as much as he could. Novak laughed and thought this might be the last time he would see clearly through both of his eyes. Hell, this guy might snap his neck, and this could be his last day on earth. Deklan wandered into the locker rooms to figure out what he needed to do next. There wasn't much to do but sit and wait. Novak changed into his fight gear and found an open corner of the warmup mat that gave him a good vantage point of every fighter in the room. He plopped down and waited for Tommy Radcliffe.

Finally, his opponent walked into the room. Deklan's entire body and mind exploded with a feeling so intense he could only describe it as hatred. He had thought about wanting to beat this guy since the fight was scheduled, so much so that he developed a hatred for a man he had never met before. Deklan had convinced himself that his opponent wished him dead in the ring and that drove him to be prepared for this in a way that he had never been prepared for something before. Novak quickly reflected on how absurd it was that one person brought into his world such intense motivation, drive, emotion, fear, excitement, and such an intense desire to win. Tommy Radcliffe entered the back fighter waiting area and made eye contact with his novice opponent. Novak could not stop the biggest smile from forming on his face. There was no turning back now. Novak smiled at Tommy Radcliffe. Without so much as a twitch from Radcliffe, Deklan knew he had rattled his opponent. That may have been a big mistake.

Most of the fights had already finished. Two more to go before Deklan hit the canvas. The arena was at full capacity and every person in the house was fully charged. Novak sat in a dark stairwell behind

CHAPTER THREE

some curtains with his gloves on and his mouthpiece in, listening to the roars of the crowd. Novak was as warmed up as he could get. Every cell and muscle in his body was electric and bursting with anticipation. Novak heard the crowd erupt when the fight preceding his own fight ended by knockout in the first round. Deklan fluttered between terrified and confident. Minutes stood between his debut and showtime. In that moment everything became still, and the great sounds grew muffled. Novak felt the blood rush to his head accompanied by a buzzing that seemed to drown out everything else. Deklan thought he was going to pass out but instead he fell into a strange and deep sense of well-being, and he entered a serene space that felt uncharacteristically tranquil. Novak noticed the calm in the moment, but rather than question it out of the fear of losing it, he just accepted the sensation and accepted the sudden peaceful yet wildly alert state. Deklan looked around and felt like everything was in slow motion. The yelling voices from the arena suddenly were distorted, muffled and distant. The crowd became no more than a gentle background hum, gradually reduced into a soundless state of anticipation. Deklan's body felt light, like he might need to be tethered to the ground so as not to hit his head on the grand auditorium ceiling. If he was not about to enter a death match, he would have welcomed the detached state of being.

The euphoria wasn't going to last long, "You are up, Mohawk!", someone pointed to Novak. Along with a reality jolting smack on the back, the alert came again.

"Let's go! You're up, Geronimo! Get out there." Back of the house people at MMA fight venues absorbed much of the bravado associated with the combatants.

The feel-good fantasy was over. The shit was about to become real. Turning the corner and stepping out from behind the backstage curtain, Deklan walked into the lights. The crowd was still very much there and the roar of the crowd came rushing back at him like a tsunami. The metal octagon cage suddenly resembled the Flavian Amphitheatre in the center of Rome, Italy. The gladiators were assembled. The VIP seating around the cage was vibrant, if not agitated. The drunk and rabid spectators were screaming and yelling inches from the cage. Novak recalled the intense yet distant and delirious blood lust in the eyes of the spectators. They shamelessly looked him up and down inspecting every inch of Deklan's 195-pound chiseled to perfection fighter's body. Novak knew that he looked good, and that he looked like an MMA fighter. In just a few moments he would find out if he was one.

Nobody knew who Deklan Novak was. Nobody knew his name. The fighter across from him had a 4-0 professional record. And he was on his home turf. These were his people. But when the crowd saw Novak…veins popping, body pulsating, mohawk haircut…they could feel the determination in his eyes, and everything shifted. Deklan Novak made eye contact with a heavy-set drunk in the front row of the VIP seating. Beer sloshing around in his clear plastic cup, Deklan was close enough to hear him as he leaned over to his friend and said:

"I'll bet you $100 straight-up that green short guy wins." Novak wore green shorts.

"Hell yes. Radcliffe is 4-0 with four tap outs. One was in the first round! Your dude has never fought before." His friend replied.

Novak smiled at the heavy-set drunk and gave him a subtle nod of the head, He immediately replied to his obnoxious friend, "Make it $200."

The bell rang, "Let's get it on." Announced the referee with a wave of his hand between the two fighters. They bumped fists in the usual prefight respectful greeting. Novak's opponent advanced quickly. Deklan wasn't about to back down but also did not want to get into a tactical footwork game with a professional boxer. Novak had built a strong reputation as a stout grappler and ferocious ground and pound menace. But no one that trained with Deklan Novak feared his stand-up striking game. Deklan Novak certainly had heavy hands with knockout power and a very tough chin as would be verified in short order. Skill and finesse in the stand-up fight game took many years to acquire. Novak had not even begun to develop his standup game. The hole in his fight game was no mystery to the brothers who trained Deklan and who were now yelling as his corner men to change levels

CHAPTER THREE

and to take the man to the ground. Although Deklan agreed to the game plan to immediately close the distance and take the fight to the ground, he really wanted to see and feel the speed and power of a real striker. Novak threw one jab. Then with a slight left stab step Deklan decided to launch his first combo of left jab then right cross, but before the combination became airborne, Novak's opponent landed a sledgehammer right straight punch to the bridge of Deklan's nose and followed the right hand with a violent counter jab. Novak had never been hit like that and it weakened his knees as he stumbled back a half step.

Pain shot from the bow of his nose to the back of his spine and Deklan felt a stream of warmth run down the front of his face. Fifteen seconds into his first fight and his nose was broken and he was bleeding like Jerry Quarry vs Muhammad Ali. Deklan knew he was in trouble and Tommy Radcliffe knew it as well. Radcliffe pushed the attack. Deklan's instincts brought him back to his strength and as Radcliffe came in with a flurry of striking attacks, Novak was able to find a way through the flurry to slip under the punches and into a body lock around his waist. With the gap required for effective striking between the two men closed, Novak was able to manipulate Radcliffe's back into the cage. Despite a few viscous body shots, Novak was able to regroup. "Stupid me", Deklan thought, "Now I have to fight without being able to breathe out of my nose."

Novak kept maximum pressure on Radcliffe's solar plexus as he whittled away at his posture slowly working him into the corner and finally down to the floor on the mat. Radcliffe would manage to escape, and Novak would put him back into the cage for the rest of the first round. Deklan Novak was a bloody mess when he sat down in his corner. The crowd was maniacal. "Holy Shit do they love blood." Novak thought. The Russell brothers were cool as could be. The brothers immediately began breaking down what needed to happen in the next five minutes if Novak did not want to bleed out before the fight was over. Dean plugged Novak's streaming nose and wiped him clean while he searched for any other major open wounds. Coagulating blood often hid further damage to a fighter that was not visible.

Novak took one sip of water and what felt like two short breaths of air and the referee was calling the fighters back to the center of the ring. Novak's lungs were already screaming but he refused to show any sign of fatigue. Deklan popped up and punched together his fists and reminded himself what he was there for.

Pretending to desire a stand-up boxing match, the Jui Jitsu training Deklan had kicked in and allowed Deklan to duck under the overzealous striker for a takedown. Feigning the upright combat was a well-trodden path for good reason, it worked well, and it worked repeatedly. Deklan was able to lift and elevate his opponent. One thing a raucous crowd loved as much as a spinning back fist was air mailing a guy four feet over his head for a roof rattling slam to the center of the mat. Radcliffe had nowhere to go now. Freshened up, quicker, and stronger than he was at the end of the first round, the combatants were playing Deklan's game in the second round. No more pretending. Deklan was determined to do what he knew how to do. Novak ground Radcliffe into the canvas making his 195 lb. frame feel like a Mack truck working its way through the center of Radcliffe's rib cage. The relentless assault by Deklan compressed Radcliffe's lungs like a polar bear standing on his chest. The second round was Novak's. Deklan dragged Radcliffe around the octagon staining every square foot of the mat with a mixture of his own blood and the blood of his opponent. During the second round, most of the blood in the octagon was Radcliffe's. Deklan's nasal levy had broken by the end of the second round however, and there were several large gushing openings above and around Novak's eyes and nose. With the ringing of the second round bell the entire building knew they were watching the fight of the night.

The Russell brothers were laughing hysterically while Kirk Russell plugged Deklan's nose up once more.

"Well, I told you so." Kirk said with a gleam in his eye. "You lost the first round for sure, but you probably won the second round," He critiqued in his usual honest kindly manner. "I suggest you submit him this round just to be sure," Kirk Russell laughed again, looked at his brother, looked back at Deklan Novak, and smiled coyly like

Chuck Liddell did when he was rising from his stool to finish an opponent in the fifth round. "Get out in the octagon, man, and stick to the game plan. You're going to win this fight if you don't do anything stupid."

"What the fuck?" Novak mumbled between the constant gasps for oxygen.

"Go!" The bell rang for round three. Fuck, Novak thought, those rest periods between rounds were way too fucking short.

How did Radcliffe suddenly learn how to defend takedowns, Novak thought as they started the last round. Novak had found another angle he had not unveiled in the first two rounds and took a shot on Radcliffe's legs to start the third round scoring off with a takedown. This time Radcliffe sprawled hard popping his hip deep into Novak's shoulder and striking him several times in the back of the head and neck. He stymied Novak's attack like Dan Gable did to Deklan Novak a few years before at his wrestling camp at the University of Iowa while demonstrating double leg take down defenses. Dan Gable was universally considered the greatest wrestler of all time. Dan Gable was a two-time NCAA Division 1 National Champion, a world gold-medalist, an Olympic gold medalist and a member of the United World Wrestling Hall of Fame and a member of the International Sports Hall of Fame. President Donald Trump awarded Dan Gable with the Presidential Medal of Freedom. Deklan Novak spent two weeks training with the Michael Jordan of wrestling when Deklan was in high school. The two weeks spent on the Iowa campus left an indelible mark on Deklan Novak, not so much for the wrestling skills acquired during the camp, although those new skills were incredible, but for the mental approach to combat that had a profound effect on Deklan Novak for the remainder of his life. Dan Gable told Deklan Novak that until defeat is simply never an option, you will always remain vulnerable. There was never a single Navy SEAL awarded his Trident that did not possess what Dan Gable told Novak at his camp.

Coming out of Radcliffe's double-leg takedown escape to start the third round, Novak was able to land a few punches to Radcliffe's midsection which opened him up to allow Deklan access to a clinch fighting position. Here Novak worked his opponent back against the cage unleashing an avalanche of short uppercuts and hooks followed by solidly connecting knees to Radcliffe's chin and sternum. Novak felt the air in Radcliffe's lungs purge violently out of his system with every strike he carried out. Deklan knew he couldn't slow down. He needed to finish the fight.

The drunk VIP was now directly in front of Novak, his lunatic face no more than six inches from the two fighters in the cage, Novak could smell the booze on the spectator's breath as he screamed and spit all over the cage.

"Finish him! Goddammit! Fucking kill him!" The drunk was yelling.

At that moment Novak took an elbow from his opponent sending blood, spit, and sweat flying through the air directly into the face of the VIP drunk and his friends on either side. The chain link Octagon fence did little to block the mucus and perspiration spray from the fighters. Hiding himself from further blows inside the safety of the clinch, Deklan watched the spectator's initial reaction of disgust turn into the Jack Nicholson character in the film, Wolf. The moon and the blood turned Jack Nicholson into a werewolf. The blood and the sweat from Deklan and Tommy Radcliffe turned the fat drunk and his friends into octagon participants inside of their own minds. The group yelled and screamed with their arms raised. They were flailing about next to the ring with spit, beer, and the blood from the fighters all over their shirts and bloated faces. The spectacle fired up Novak. Combat sports were not sophisticated, but the challenges significantly raised the testosterone levels of the fighters to unknown levels. To see what needed to be done, Novak remembered that he had to win the round, and the round win needed to be decisive. With two minutes left in the third round Novak could not leave the decision up to the judges.

Deklan took the crown of his forehead and began driving it into Radcliffe's larynx while sliding up under his mandible straightening him out and opening Radcliffe up long enough to slip down into a double leg takedown against the fence. Tommy Radcliffe's feet were wobbly. He was tired and his feet slid out from under him. Radcliffe hit the mat hard. Deklan unloaded most of the last of his treasure trove of stored energy and ferocity long enough and landing enough to force Tommy Radcliffe to make the mistake of lunging his head and arms towards Novak. Novak used the momentary mistake by Radcliffe to quickly and cleanly lock up a front headlock. The headlock was tight, and Radcliffe was very tired but so was Novak. Deklan moved to finish the submission but as he applied the pressure around Radcliffe's carotid arteries with his bicep and forearm, Tommy Radcliffe slipped a hand in between Novak's arm and Radcliffe's neck. The hand allowed Radcliffe to slip out of what appeared to be a tap out. The final round ended.

The fighters collapsed together in mutual respect and mutual exhaustion. Three five-minute rounds in an MMA octagon felt like running a marathon inside a twenty-six-mile rugby scrum. Team sports brought on a collective exultation, not comparable to what Deklan felt at the conclusion of his first fight. Losing at Ironman in Hawaii was not finishing the race. Losing on Mt. Denali was not reaching the summit. Although there have been many more deaths over the decades on Denali and Everest than in the ring, the reality of the battle in an MMA octagon or a boxing ring was centered on an opponent solely present to hurt another human being. Mike Tyson once summed up his intentions within a boxing ring when first eyeing his opponent... *"I want to break his will. I want to take his manhood. I want to rip out his heart and show it to him."* Deklan's hand was raised as the winner.

Competing in the octagon involved absorbing a back fist to the face, absorbing straight punches to the face, neck and solar plexus. Fighters took full-speed knee strikes to the face and abdomen, endured locked arm bars designed to fracture limbs, rear naked choke holds designed to cause blackouts, Guillotine Choke holds, Triangle Choke holds, Anaconda Choke holds, Achilles Locks, and Kimura Joint locks taught from Brazilian Jui Jitsu masters that were designed to snap elbows and shoulders. Deklan recalled years later while assessing the moments following his first MMA fight, a terrifying documentary he had recently watched called *Free Solo*. In June 2017, Alex Honnold after his successful free solo climb up the granite face of El Capitan in Yosemite would have experienced a similar exhilaration to an MMA fighter after a victorious fight because the climber stared down death and never flinched. Directed by Jimmy Chin in 2018, *Free Solo* not only captured the physical challenges but also the unique mindset necessary for Honnold to succeed. Recalling his first fight, Deklan exhaled while he gauged the comparison. It was fair.

Deklan didn't smile, he didn't celebrate, and he didn't acknowledge his new group of superfans. The fighters embraced one more time and Deklan went straight back and sat in the stairwell that he sat on twenty minutes earlier before his nose was broken. He found solace on the stairwell he occupied before his lip was split open in two places, before his brow was spit like an open sardine can above the right eye, before the cartilage in both ears began swelling quickly, before the bruises and cuts and scrapes were everywhere. Deklan just sat in exultation, staring directly through the floorboards into the center of the earth. Novak had never felt exhaustion like that. He had never felt ecstasy like that. Novak had never felt pain like that. Most of all, Deklan had never felt contentment like that.

Novak and Radcliffe were awarded the fight of the night for their heroic efforts. A few more dollars were added to the evening's paychecks. That evening Novak drove himself home alone and did not shower before he left. Deklan walked into his mom's house through the back basement door. Deklan had told no one at his home that he was fighting professionally for the first time that night. Deklan showered and laid in bed, alone with his triumph. Novak slept more soundly and deeply than he ever had after his first professional fight. Deklan Novak faced another man toe to toe who wanted to put him in the hospital or worse. In Pulp Fiction, Butch Coolidge (Bruce Willis) was asked if he felt bad about the other fighter who died in the ring during their recently completed fight. Butch, who was unaware of the fighter's demise, replied to the curious female cab driver,

Esmeralda Villalobos... "I don't feel bad at all. If he had been a better fighter, he would still be alive."

One other thing became increasingly clear to Deklan that night and each night after he won another fight. Deklan built up a 7-0 record in MMA fights following his first fight with Tommy Radcliffe. People in the sport of mixed martial arts around Chicago were taking notice of Deklan Novak. The MMA was not the path for Deklan's career. The MMA and the 7-0 start to his professional fighter career gave Deklan Novak the confidence that he could become a Navy SEAL. Eight out of ten men failed to become a Navy SEALs when they signed up for the training. Novak proved to himself that he was one of the 20% that could make it through BUD/S. Deklan Novak was about to join the Navy.

CHAPTER FOUR

2010... BREACHER'S SCHOOL. NSW BREACHER SCHOOL...SOMEWHERE ON THE EAST COAST

AFTER RETURNING FROM AFGHANISTAN, Deklan was certain that he wanted to be a breacher. During the 2009 deployment, Deklan's missions had assaulted several targets that required complex planning and execution from the group's lead breacher to gain entry to the target compound. Novak had become friends with the DEVGRU (SEAL Team Six) lead breacher, E-6 (PO1) John David Kingsbury (JD). The lead breacher was just a couple of years older than Deklan, and JD was a beast on the wrestling mats in the MMA room which was set up inside of the Army's Special Operations Camp Vance at Bagram Airfield in Afghanistan. There were several DEVGRU guys, Deklan, and a few of the ST1 (SEAL Team One) Alpha platoon that liked to spend time on the mats and train MMA. This lead breacher was one of the best MMA fighters Deklan had ever seen. Kingsbury eyeballed Novak one day and waved Deklan out onto the mat with him and Deklan gladly accepted. They were buddies the moment they slapped hands. The pair cranked on each other's head, neck and limbs for an hour and a half. The DEVGRU breacher invited Novak to do a workout with him and a few guys afterwards and then to grab lunch chow with them. That evening JD requested that Deklan Novak join him during that afternoon's mission planning, pre-ops, and to shadow him on the night's operation.

It was exciting to watch for Deklan, to say the least. Kingsbury scoured all the intelligence and available photographs of the target compound. He made half-a-dozen phone calls to the intelligence collection guys who then inquired with their informants. PO1 Kingsbury asked dozens of questions and when only some of the answers came back, he improvised by scrapping a plan of entry and started a new plan with remarkable innovation and speed. The entire operation was dependent upon the breacher's assessment and decisions of how the target should be approached and breached. The targeted compound was in the middle of nowhere. Ten buildings were surrounded by an eight-foot-high wall surrounded by two miles of farmland surrounded by 7,000-foot-tall mountains. So whatever entry point on the compound's wall the lead breacher determined was the best would determine the cardinal direction the mission would infill and approach from. The breacher's assessment would determine the insertion point from ten clicks away. The preferred insertion point would determine the route of approach by the team's helicopter transport support. The 160th Special Operations Aviation Regiment (Airborne), also known as 160th SOAR or the Night Stalkers, provided helicopter aviation support for special operations forces, including DEVGRU in Afghanistan. Their missions included attack, assault, and reconnaissance usually at night, at high speeds, low altitudes, and on short notice. So much of the planning for getting to the target hinged around the breacher's decision of how best to enter the target compound. The point-of-entry for the compound determined all the close quarters combat directives once inside and roof top over watch positions were determined by the breachers point-of-entry.

John David Kingsbury would say, "There are three basic principles of combat; shoot, move, and communicate. If the breach fails, movement ceases, the assault force will not make entry, and the mission comes to a halt. The consequence of failure is the possible loss of lives and national prestige. Being able to enter this facility is a key element in the success of any assault mission."

Once inside the target the breacher never ceased working problems. Not only was he a viably efficient operator at close quarters but he was a brilliant tactician and problem solver. One moment he is called from across the compound to a double locked back door of one of the outbuildings where he quickly slips his collapsible cable cutter from his back, snips both the locks, they fall silently into his hand, and he sends the entry team inside the swinging open door. The next moment the breacher may be called back across the compound operating autonomously in the uncertain environment. JD often had Novak running around with him while scanning the doorway and surrounding walls for any explosive traps, Deklan's mentor determined that one of his patented donkey kicks would spring the door wide open. And it did, like a choreographed training video sequence.

Deklan Novak was certain that the energy and dynamics of the Breacher were right for the ST1 single pump. Novak put in the request to become a breacher first thing when he got back to San Diego from Afghanistan. The rule of thumb in the SEAL Teams is that after the first deployment if everything went well, the team member was allowed a request to attend the specialized school of choice. SEAL Team members became specialized in one primary area. The options were NSW Communications School, NSW Lead Breacher Advanced Close Quarter Combat, NSW SEAL Sniper Course, and NSW Special Operations Combat Medic Course. In addition, there were dozens of other courses available for SEALs to attend over the years.

And there were many more. With Deklan's best friend on ST1, Frank Carlin, they requested Breacher and were both granted their first preference. In two weeks, they were flying east to attend the summer session Lead Breacher School.

NSW Lead Breacher and Advanced Close Quarters Combat school was three grueling weeks at an undisclosed East coast Army facility flap jack in the middle of nowhere and during the thick humidity of the summer of 2010. Frank Carlin and Novak got to work early. They were the two youngest students of the 33 SEALs from around the country. The other SEAL students at Breacher School had all varying levels of experience. One SEAL was a 37-year-old chief, another had five previous combat deployments, several had multiple deployments as Marine Reconnaissance before switching to Navy blue and becoming SEALs. Five or six fellas had been in the Navy fleet for ten years before making it as a SEAL, so they already had 15 years in the Navy. Carlin and Deklan were what was called one pump chumps meaning they only had a single deployment (a single pump). Being light on experience didn't matter. Carlin and Novak were hungry like starving bobcats.

Deklan was so ready to go that on day three Novak had decided that he would quietly become Honor Man of the current breacher course. Novak knew it would not be easy since there were several very proficient and highly skilled experienced SEALs at the school. Also, the older experienced guys had all taken up the high-profile leadership roles. These were the positions that garnered the most attention from the instructors and fellow peers. Honor Man recognition would be weighted between both instructor votes and peer evaluation. Novak would have to not only make a mark as an individual, but he would need to become known as the most indispensable cog as a fellow teammate and peer. To make it even more of an interesting uphill battle, there had never been a young "new" guy that was able to earn Honor Man in the long history of the school. Nothing was going to deter Deklan Novak from becoming Honor Man. Without a word to anyone, Novak dedicated the next three weeks to dominating every evolution in every way possible. Deklan would take no prisoners during his quest and the decision would not propel his popularity within the breacher school. Hot shot aggression was often envied and always resented.

The NSW Lead Breacher and Advanced Close Quarters Combat School students learned how to conduct breacher mission planning and target analysis. Breacher students learned to perform as a lead breacher using manual entry tools, ballistic entry tactics techniques, mechanical entry techniques, exothermic entry techniques, and explosive entry techniques. More follow-on procedures included directing

CHAPTER FOUR

entry team positioning and security in tactical assault scenarios. The Lead Breacher performed manual entry techniques utilizing sledgehammers, light and heavy weight rams, hooligan tools, bolt cutters, pry bars, automatic spring punch, spreader bar, and a fire axe. The breacher learned how to perform mechanical entry tactics utilizing chainsaws and quickie saws. The Lead Breacher learned target entry via melting or vaporizing objects, locks, and metal bulkheads using exothermic torches. He learned to ballistic breach utilizing the quick and effective means of shotgun breaching for gaining entry through light to medium doors. The Lead Breacher became proficient in explosive breaching utilizing a wide array of precise and specific types of explosive charges for difficult and fortified forms of entry.

The Course was taught by five well-trained active-duty SEAL Lead Breachers from around the country. Two had recently returned from deployment in Iraq with a west coast SEAL Team and were going to spend the next 18 months shore-side as instructors recovering from a few combat injuries. Two of the instructors were ex DEVGRU operators and breachers and were instituting new training techniques that they had been developing and employing for the last few years as DEVGRU operators in Afghanistan and Iraq. The Course Master Chief was a 30-year SEAL and 25-year Lead Breacher that had been a member of 6 different SEAL teams and deployed 11 times. In addition, 3 or 4 civilian contractors with varying degrees of military and law enforcement breaching and explosive experience trickled in and out of the classroom and training facility sharing technique and sprinkling bits of wisdom and experience.

Novak's favorite instructor made it a habit to remind the students every morning that the ultimate objective of breaching could be described as, "the most efficient use of the minimum amount of explosive, in order to achieve the 100% penetration of the desired target, 100% of the time."

The instructors had a knack for making concepts clear by reciting history as part of the lesson book. They explained that early explosive entries were developed for hostage rescues and situations where armed suspects were believed to be barricaded.

One of the first documented explosive breaches was done by the Los Angeles Police Department in December 1969. Barricaded militants shot three officers. Entry was made by a large explosive charge placed on the roof of the barricaded structure. After detonation, there was a large hole, and major damage was accomplished on the building. To the surprise of everyone present there were no injuries. In the end, the heavily armed suspects surrendered. Lessons were learned.

In the early 1970's, the Navy's Explosive Ordinance Disposal (EOD) units, while training with various civilian law enforcement bomb squads, were first introduced to explosive entry techniques. EOD units conducted testing throughout the 70's and 80's to improve their breaching techniques.

The EOD Technology Center (EODTECHCEN) conducted the first documented breaching tests in 1977 at Indian Head, Maryland. The test subject was "Rapid Entry into Locked Buildings Using Explosive Devices."

In 1983, tests were conducted to investigate the feasibility of using explosives to gain rapid entry into rooms containing improvised explosive devices, (IEDs), where hostages might be involved. Test results in the collection, for the first time, showed significant amounts of data concerning air blast and cutting effectiveness of various types of explosive charges for breaching. Many units today, Army, Navy, USMC, and most large police departments, use explosive breaching to enter facilities.

The best training was always the actual hands-on experience. There was no substitute for field experience. The user must go through the classroom to understand the theory of explosives to apply breaching at its full potential. Breaching consisted largely of on-the-job training. It is an ever-growing process. Complacency is the greatest potential cause of accidents in breaching. There was no room for short cuts when working with explosives. SEALs were taught to always be professional, to get involved, and to test one's inner limits. Deklan's only limitation was his own imagination.

Frank Carlin was one of those guys that lived a charmed life. Everything was easy for him. He was stronger, faster, and smarter than anyone in the room. And Novak was certain that Carlin knew it. Carlin was laid back and chilled out right up to the point that he needed to begin exerting himself. He did not waste any energy; He did not try that hard nor care very much and yet he always finished in the top 10%. Deklan, on the other hand, always felt like he could do more or could do better during training, on the mat, or in the field with an almost insatiable appetite to be competitive at the highest levels. Often in BUDs training, Carlin and Novak were paired up. Regardless of how far ahead of the rest of the pack they were, and they were almost always in the lead, Novak wanted to extend it. Carlin would look back at Novak and ask why are you pushing this so hard? We are five minutes ahead. Novak would make the argument that he wanted them to be the greatest pair of land navigators that the course had ever seen. Deklan Novak wanted to break every BUDs record. For example, on the two-day Alaska training course, Novak announced that he didn't care if there was a fresh 14 inches of snow, he knew that they could still finish hours ahead of the next pair. Carlin looked back at his obsessed teammate, laughed and shrugged his shoulders. Carlin pushed ahead, harder. Together, they broke training record after training record.

Novak hesitated several days to tell Carlin about his Honor Man goal until one morning drive in from the hotel. Frank Carlin was driving their white Toyota Camry rental car from the Holiday Inn off the nearest exit from the Army base. Merging onto the highway he looked over at Novak with a grin and asked if he thought that he had a shot for Honor Man. They laughed. Clearly, Carlin knew what Novak was up to already. Carlin said he thought that Novak had as good a chance as anyone. Deklan asked if he wanted to try for Honor Man. Frank Carlin smiled, laughed, and said, hell no.

By the second week Deklan had made a name for himself. He found himself in many one-on-one conferences with the instructor staff often laughing and asking questions about his life. One manual breaching exercise was where each student was required to breach a series of four barricaded doors using only a sledgehammer, Deklan finished all four with all the intensity of a caged demon. Novak didn't realize that all the instructors had stopped to watch as he violently smashed through each of their well-designed barricades with ease to finish the exercise before some guys had cleared the first barricade. After another exercise requiring some skill and a lot of brute force, one of the DEVGRU instructors approached Deklan and asked, "Is everything ok? Are you angry about anything? Do you need to talk to someone?" he asked again with a smile. Novak's chest was still heaving, and the muscles and veins were still attempting to surge their way out of his forearms. Deklan assured the instructor that while intensity had never been an elusive state, there was no anger management classes needed and that he simply loved smashing things into pieces.

Novak knew going into the final five days (testing phase) of the course that he had the support of most of the instructors. They loved Deklan's intensity and seeming recklessness and awesome ability to breach anything. But Deklan did not know if he had the overall peer evaluation. Novak thought that his over-the-top hard work annoyed and pissed-off a lot of guys. He knew several guys were just plain jealous that he could physically perform to such an exaggerated level all day, every day. Additionally, resentment grew from his peers because Novak was smart enough to easily finish and pass every written test with ease. Novak had a plan to combat the resentment or jealousy or whatever his high-octane motion caused. As soon as they assigned the teams during the testing phase, Novak volunteered for as many leadership positions as possible. Deklan decided that he needed to show the group that he could both kick ass and lead. Deklan also knew that the process would provide the opportunity for him to give all the accolades to the guys. Deklan was going to make every guy on the team look better. Novak set out to make his success reflect a team effort and each member of the team would feel what he felt daily. Novak would make each team member feel like a god.

Deklan had an entire team of eight young gods, sculpted warriors in search of skills so unique they altered warfare…so the groundwork for lifting aspirational pedestals was easy. Deklan scoured all the strengths and weaknesses within the team and began putting each member in the places that would allow the team to conquer any chal-

CHAPTER FOUR

lenge presented to them. The last three days of Lead Breacher School were nonstop practical competitions between Deklan's eight-man team and the other three groups of eight and seven-man teams. Two guys had failed out early during the second week of training. The LPO of one of the other groups was a ten-year prior fleet enlisted guy. That meant that he had done 10 years in the regular Navy before he entered BUDs and became a SEAL. He also had just finished his second deployment with SEAL Team Two and he clearly had plans of going to SEAL Team Six at the earliest possible opportunity. He was good and as far as Novak could tell he was the most likely candidate for Honor Man. Although there were several other strong candidates to include the leaders of the other two teams, he, and Novak, appeared to be the front runners. Novak never let off the gas during those last three seemingly continuous days. No matter how tired his guys got, Novak worked and worked to keep them motivated and performing well. They excelled at every task and did great. As a team they were far ahead of the other team competing with Deklan for Honor Man. Deklan needed to make sure that there was no way he was not the obvious choice for class Honor Man.

Entering the final day of Field Training Exercise (FTX), Deklan had to make a choice if he wanted to stay as team lead which would mean that he would not be operating as a breacher on actual breach problems. As the team leader he would be coordinating where and how his guys would be placed to address and clear the over 30 barricades and breach problems they would encounter in the kill house that evening. Or Novak could allow one of the other older guys to run the team and he could take a place as one of the lead breachers. When it came time to make that decision the guys on the team unanimously asked Novak to stay on as the Officer-in-Charge of the team. Deklan did not hesitate and took the OIC role with honor and they got after it.

After the instructors and staff reset the barricades, Novak's team was the second group to enter the breach house. The clock started when the initial seven-foot strip charge was detonated to tear the hinges off the front security door of the building allowing initial entry. The seven-foot strip charge was primarily used as an external charge on outward or inward opening heavy wood or metal fire doors. It was often used to gain entry through light masonry walls and windows. The charge was simple and very reliable, compact and could be rolled small to endure hard insertions. It was very versatile and could be used on a variety of targets and could be adapted to any size. The seven-foot strip charge was constructed of one strip of explosive Deta sheet cut into a one-inch by seven-foot strip. A split MK 140 booster, two 16" sections of detonating cord for pigtails, and a Dual NONEL firing system made up the initiation system.

Their initial entry was flawless, and the eight-man team entered the facility without a problem, and they began working through barricaded doorway after barricaded doorway. Often, they needed to utilize five or six different breaching methods on a single entry point due to the way it was barricaded. For instance, a door would be wrapped in metal wire and rebar requiring the use of a quickie saw. Novak and the entire team had been hammered for nearly three weeks on the consequences of a breacher failure. Teammates would die if the breacher failed to do his job. Beneath the metal wire and rebar were multiple 2 x 4's requiring a chainsaw. Behind the door, the team ran into nylon rope and cloth, unexpectedly jamming the gears of the chainsaw. Novak directed the team to utilize the work of manual tools such as a sledgehammer and hooligan tool to fully breach the doorway. As the unexpected nylon rope and cloth barrier was untethered, three other breach problems were occurring in other parts of the pitch-black building.

The FTX culminated when the team reached the overhead steel ship bulkhead door requiring a breacher to cut overhead through the thick expanse of steel with the exothermic torch. Sparks were flying all over the room and red-hot molten lava was dripping from the tip of the torch. Deklan watched Tommy painstakingly inch his way towards a complete circle large enough to allow the team to squeeze through. Tommy was a spectacular breacher and completed the task in record time. He began yelling moments before the final bit of steel was turned to lava for someone to get over there and prepare to

CHAPTER FOUR

handle the steel disc as it was released from the ceiling and before the disc could get to the floor with an explosive collision of clamor that would compromise the team and the mission. Deklan knew the team was seconds from finishing breacher school.

"Time!" shouted an instructor as one of Novak's guys lowered the smoking red hot steel disc to the floor. "One hour and 33 minutes! Great job fellas!" It was a lot faster than the first group but there were still two teams left to finish. All the other groups finished with times close to three hours. After the clean-up Deklan and Carlin made it back to their hotel at 4:30 am with enough time to shower and close their eyes for a few hours.

After the abbreviated graduation ceremony, and flush with a fulfillment that came from a job well done, Novak's team had clearly risen to the task. The Navy Special Warfare (NSW) Lead Breacher and Advanced Close Quarters Combat School Master Chief pulled out the Ak-47 style shotgun, a Russian made semi-automatic Saiga 12, worth almost $2,000, from its box explaining that this would be the gift for the class Honor Man.

"Deklan Novak, come on up here and collect your prize!"

CHAPTER FIVE

2013 CAMP KENOSHA...THE UNIVERSITY OF WISCONSIN – PARKSIDE CAMPUS & NORTHWESTERN UNIVERSITY SUMMER FOOTBALL CAMP PART TWO. THE ROOKIE DRILLS AND SHEDDING LAYERS

NOTHING UNUSUAL OCCURRED during the first week of fall football camp with Northwestern for Deklan Novak. Nothing unusual unless putting on pads for the first time in nearly fifteen years was abnormal. Deklan Novak and 120 guys gathered, laughed, and bantered ceaselessly inside the old basketball court at UWP that had been converted temporarily into the player's locker room for the two weeks of fall camp. There were cliques of all kinds. Some by race, some by position group, some by class, and some by religion. Most of the cliques overlapped making the seams virtually undefinable. Everyone knew where they mostly belonged, and the crossovers occurred more often with the veteran team members. Whenever a good "hazing" ritual arose, all seams vanished, and the room became filled with a unifying thunder. The rumble of excitement was amped up high because 119 players were thrilled the attention was not on them.

The first time the attention was directed at Deklan Novak, many were apprehensive on more than one level. Deklan had heard a few rumors about the Navy SEAL and that no one was immune from the "hazing" at UWP during their first camp. Compared to what Deklan had gone through at BUD/S as a Navy SEAL candidate, Deklan Novak thought whatever was coming was going to be interesting and fun. Just a few cubicles from where Deklan's makeshift locker sat, hanging from the ceiling, were two old climbing ropes. Deklan's interactions with his teammates one-on-one or in small groups up until this point had been awkward and truncated. Deklan knew that talking to an active-duty Navy SEAL about teen-aged locker room, towel-swatting, dick-measuring idiocy was more than a little intimidating. On this day, the second post practice gathering after the sweltering Kenosha practice, the makeshift locker room was alive in every way. Everywhere, sweaty, gigantic human beings inside the non-air-conditioned gymnasium huddled in conspiratorial circles. Dripping wet behemoths were running back and forth, waving their tree-trunk arms, and conspiring to find the messenger with enough courage to approach the Navy SEAL. The smell of old sweaty gear, prewrapped tape, and exposed jock straps prevailed. The distant scent of lunch cooking could barely be detected. There seemed to be at least 50 conversations going on at the same time. Laughter erupted and died quickly.

Suddenly, the room became noticeably still. Deklan noticed the player next to him in the locker room developed a mischievous yet very uncertain smile across his face. The player next to Deklan Novak was scared to death. It was obvious that whatever was occurring was being directed at the only Navy SEAL in the room. A chant began forming in the room much like the chant at a football game "Let's go defense." The chant on this hot August day was "naked rope climb," "naked rope climb." Deklan Novak was sitting in his underwear with 119 other half-naked men surrounding him and they all began to join the chant..." naked rope climb." The chants evolved quickly into "Novak naked rope climb, Novak naked rope climb." Deklan looked again at the player seated closest to him and the young man was now completely terrified about how an active-duty Navy SEAL, combat veteran and former MMA fighter was

CHAPTER FIVE

going to respond to being the subject of a semi-humiliating hazing request.

Novak soaked up the attention. Deklan Novak loved every minute of the ritual and was trying to think of ways to make the moment even more epic. Novak considered a fake rejection of the idea just to gauge what the group had planned to do if he refused to comply. Deklan was certain that they had not thought that far in advance and Novak decided to comply willingly and happily. Deklan knew his rope climbing ability was spectacular and while saying no to the rope climbing request would have added to the excitement, Deklan Novak dropped his underwear and shot up the climbing rope like he was born in the jungle. Novak's arms were eating up the rope with a smooth, powerful rhythmic cadence that stunned his tormentors. The Navy Frogman was bursting towards the ceiling in effortless fashion using a technique called the L-Sit position. Navy SEALs were required to climb ropes without using their legs, often using two ropes side by side without using their legs. Rope climbs were supposed to be arduous struggles to gain minor upward advances, not jaw-dropping speed runs completed in under a minute. Deklan knew this was the first opportunity he had to officially be welcomed into the group. The hazing ritual was brilliantly used to shed the layers that needed to be shed. The naked rope climb was performed with super-human efficiency, but the ritual made Deklan Novak human to the team. Deklan was now a teammate first to the group before he was a Navy SEAL.

Eleven years as a Navy SEAL had built Deklan Novak into a stone-cold granite warrior. Deklan needed to train to become a football player. In a football game, there are 11 minutes of live action in a 60-minute game that covers between 50 to 80 plays. Football is a game of bursts...short and explosive. Navy SEALs are trained for endurance more than anything else. The foundation of training as a Navy SEAL is unmatched yet not ideal for the Big Ten football programs. While awaiting the start of NU summer camp, Deklan was training at a fitness academy in Highland Park, Illinois called EFT Sports Performance. Deklan Novak walked into EFT Sports Performance after being advised by David Talbert at Great Lakes Recruit Training Command that EFT had former NFL players on their staff and the best way to prepare for a football camp was to train with the real deal.

Deklan and Amber Novak shared a small house they had purchased just over the Wisconsin border near the base. The rural community was affordable but not desirable for a girl raised in Kenilworth, an affluent community on Chicago's North Shore, merely 51 miles south of their 3-bedroom, 1300 square foot Wisconsin ranch house. Amber had paid her dues and done her time. Amber watched those 51 miles become the radioactive gravitational separation within her marriage.

Amber Novak had waited for a full-time husband since Deklan Novak joined the Navy. During BUD/S, Amber moved to San Diego and a new family followed a few years later. Deployments and diapers were never covered in **The Commonsense Book of Baby and Child Care** by Dr. Benjamin Spock. Amber Novak was never given the opportunity to trust her own instincts. The Naval commitment to become a Navy SEAL is usually 8 years of total obligated service. Deklan Novak signed a four-year extension after 7 years in service without consulting his wife or his parents. Earning a Trident changed Deklan Novak by giving him a lifetime of respect inside a world dictated by a commanding officer and a chain of command that was endless. SEALs had remarkable autonomy within their teams, but the military is essentially the ultimate absence of personal autonomous determination. The military was a life based on taking orders. Deklan made the first of many impactful decisions surrounding his future without consultation.

As Deklan Novak neared the end of being told what to do, he was energized by the negativity that accompanied any suggestion that Division I football may not be something to pursue. Finally, when the end to Deklan's military commitment became visible, Deklan decided to attend college full-time and play Division I college football. Deklan Novak's family voiced a universal cry that Deklan tuned out. The overriding question from Deklan's wife and his immediate family was ...when is enough going to be enough? Deklan Novak's family wanted to know what more was there to prove? Novak heard only one voice inside his head. Deklan gathered his gym bag for a trip to EFT Sports Performance.

"Where are you going now?" Amber asked as Deklan was halfway out the front door.

"I'm going to a gym in Highland Park." Deklan answered reluctantly and prepared for what was sure to follow.

"What gym in Highland Park?" Amber asked. "There must be 100,000 square feet of workout facilities at the base, and you have the run of every single inch of that space. The all-worshipped Navy SEAL on the base can do whatever the fuck he wants to do, so why are you going elsewhere to work out?"

"Dave Talbert has worked out at EFT in Highland Park over the last year or two. They have two former NFL players on the staff. I need to work with these guys before I go to summer camp in Kenosha." Deklan dropped the gym bag. The exchange was not nearing an end.

"You asked me about getting a degree and I know that is a priority for our future. You did everything you could to get into Northwestern and nothing you did has anything to do with getting a degree. You were able to get admitted to Northwestern University, a fantastic school, and your only focus is to make the football team." Amber Novak was not wrong. "What am I supposed to think? You are not out of the Navy yet and you are committed to another timeline that will keep you away from your family as much or more than the deployment and training schedules for the teams."

"You tell me what is fair, Amber." Deklan asked. "I have spent the last eleven years being told where I had to be and when I had to be there. The college football effort will benefit both of us. A degree from Northwestern and the contacts inside a Division I football program are priceless."

"This has nothing to do with proving something else, does it?" Amber was asking rhetorical questions now. "That's been your fallback for everything. When someone lays down the doubt, Deklan Novak will prove every fucking one of them wrong! Everyone keeps asking me what are you trying to prove? I tell them that I don't know, but I do know. You asked me what is fair. If I operated under the guidelines of what is fair, I never would have gotten married or had kids. Women lose under those guidelines every time. You always carry the weapon and the full magazine with you. You are scared to death of an empty magazine. I didn't marry you because I couldn't wait to see what kind of PTA parents we would be. I harbored no illusions of Deklan Novak coaching the Kenosha Komets Midget Major Hockey Club. Deklan Novak was deployed during a war on terrorism against the United States. That generates an excellent cover for an absentee father. None of us expected a medal for waiting. We simply wanted the hat to change. Have a great workout, Deklan." Amber didn't wait for a response and left the room. Deklan left for Highland Park.

> *From the Chicago Daily-Times during summer camp at Kenosha for Northwestern University...three full pages, front full cover photo of Head Coach Lance Adams and Deklan Novak in an NU football uniform holding an American flag. The story fascinated 9.62 million people in the Chicago area and sent red flags flying at SOCOM.*

The front-page cover photo was simply titled...The Navy SEAL. The tagline below the title read: Deklan Novak is 32, married and has a full-time job. He's trying to make the Northwestern football team as a walk-on. He's also an active-duty Navy SEAL. The odds are stacked against him, and that's just the way this Navy SEAL prefers things. Ron North reports, pages 8-9. The major story went into detail how Deklan was 32 years old married with three kids and held down a full-time job on top of being a full-time NU student trying to walk-on a team as a defensive end/linebacker. Novak was competing against scholarship athletes aggressively pursued by the university. Head Coach Lance Adams described his new walk-on candidate as "relentless, nothing unexpected from a SEAL."

The story went on to detail a typical day in the life of Deklan Novak, starting at 6 a.m. strength training and ending after 8:30 p.m.

when Deklan's last class ended. Then Deklan had to drive back to the Illinois-Wisconsin border where he lived with his wife and three kids. Novak's job while he was trying to make the NU football team was as a Navy SEAL instructor at Great Lakes Naval Recruit Training Center in North Chicago. The story gave background on Novak from his BUD/S class to his assignment as a breacher for SEAL Team One. The nature of Deklan Novak's deployments was classified but the Daily-Times story mentioned blowing up targets one day to fast-roping out of a helicopter the next day to parachuting 20,000 feet for a nighttime combat insertion during Operation Enduring Freedom. Novak and the story's author agreed that the general comparisons between life as a Navy SEAL and as a Wildcat football player were inappropriate. Winning on a football field and winning in battle had no link together. The traditional war analogies used to describe sports like: they're going to war on the field, they were blown away, they went to combat in the trenches, they were throwing bombs...were inappropriate at best and aggrandizing when a combat veteran from the war on terror was the storyline. Win enough on a football field and the players have a chance to play in a bowl game. Win enough as a Navy SEAL and you earn the right to go home alive. Several of Deklan Novak's SEAL teammates did not make it home. "That's the path you take as a soldier and a warrior." Novak said. "You just learn to come to terms with what is going to happen."

The story quoted Deklan's wife, Amber. When asked about being separated from her husband for more than 280 days a year as a Navy SEAL, and now Deklan had the daunting schedule of a full-time NU student athlete to go along with his full-time job at Great Lakes Naval Base, how was she able to handle three kids and all the work necessary to raise a family.

"Our family really likes to challenge each other, and it's important to us to aspire to our dreams because I don't think that you can truly be happy unless you go after those dreams that you think are unattainable." Amber told the Daily-Times reporter. Before summer camp was over at the University of Wisconsin-Parkside for the Northwestern Wildcats, Amber Novak had taken the three boys and moved back to Kenilworth with her parents.

The major story sent tidal waves through campus and lit up the phones from the Commanding Officer at Great Lakes Recruit Training Command to Special Operations Command (SOCOM) at the Naval Amphibious Base Coronado. To put it bluntly, The Navy was not happy to find out that an active-duty Navy SEAL was playing football for any college, much less for Northwestern University, one of the most highly acclaimed and famous collegiate institutions in the United States. Master Chief Evan Kerr, Deklan Novak's Commanding Officer at Great Lakes, put in a call to locate his nomadic warrior. The Daily-Times article made the search easy. Head coach Lance Adams sent word to the team dormitory at UWP for Deklan Novak to come see the head coach in his office and the UWP campus. Practice had ended for the day. Novak showered and went to see what his head coach wanted.

'Sit down Deklan." Coach Adams motioned to Deklan Novak with his hand where to sit. "I got a call today from your commanding officer at Great Lakes. Seems like the Navy is not all too pleased that you are playing football at Northwestern. It was crystal clear from the phone call I received that the Navy was unaware of your football aspirations with regards to playing for the Northwestern Wildcats. Is that a safe assumption that I have made?"

"That is correct coach." Deklan did not hesitate. "I saved up most of my leave time for the year so I could attend summer camp without having to explain the move to my command. I know you told me that if I got into Northwestern, you would have a spot on the team for me. I needed to know if I could hang with these guys on the field. Fifteen years out of football is a long time. Less than a week into camp and I know I can play with these guys. I'm not trying to disrespect anyone out here, but competing with these young boys is one thing and seeing significant playing time is another level that I have yet to find. I have not told the Navy about trying out for the team. My guess was that they would have tried to stop me."

"You thought that the Chicago Daily-Times running a full-page front cover photo of you and a three-page article was going to assist in hiding your football career from the Navy?" Adams asked smiling sarcastically.

"No sir." Deklan replied. "I knew the article was going to call up a full-on brass meltdown, but my hope was that the good publicity for the Teams might help alleviate some of the negative fallout from California and believe me, there is a shitstorm coming my way."

"You have been told to report to Great Lakes immediately." Coach Adams informed his new player.

"I am not surprised. I got dropped off here, coach." Deklan hadn't thought about a ride until that moment.

"I'll have my assistant drive you back to the base. Good luck and keep me informed what happens there." Coach Adams called for his assistant and instructed him to take Novak back to Great Lakes Recruit Training Command."

"Hoorah, sir." Deklan Novak knew this was coming eventually.

CHAPTER SIX

MORE 2013...SUMMER PRACTICE AND THE CHICAGO DAILY-TIMES

MASTER CHIEF EVAN KERR went to bat for Deklan Novak. Master Chief Kerr was an old-school Navy SEAL from the Vietnam era, when the Viet Cong called the SEALs, "men with green faces." Established in 1962 from a Kennedy White House commitment to develop America's unconventional warfare capabilities, the present-day Navy SEALs had been mostly "frogmen" during WWII and the Korean war. Soon after the Vietnam War started, the Frogmen were carrying rifles and became experts in special operations tactics. The Navy SEAL credo was sacred to MC Evan Kerr. The Navy SEAL credo paraphrased included but was not limited to:

My loyalty to Country and Team is beyond reproach. I humbly serve as a guardian to my fellow Americans always ready to defend those who are unable to defend themselves. I do not advertise the nature of my work, nor seek recognition for my actions...

. . .

The times had changed in dramatic ways. MC Kerr had no problems with former Navy SEALs writing fiction or former Navy SEALs using their experience to pass on valuable lessons to the business community. Marcus Lutrell's book Lone Survivor was a gripping and tragic recantation of Operation Red Wings in 2005 under the command of LCDR Erik S. Kristensen of SEAL Team 10. Nineteen Americans died. Eight SEALs and eight Army Night Stalkers died from a rescue mission shot out of the sky by a rocket-propelled grenade. Matthew Axelson, Michael Murphy, and Danny Dietz died during the original four-man insertion gone bad. Many claims from the book were disputed including the claim that Ahmad Shah was "one of Osama bin Laden's closest associates and a high value target". Shah was not a member of al Qaeda, nor did he know Bin Laden. There were numerous disputed accounts of the enemy numbers during Operation Red Wings from as few as eight to as many as 50 or more. The number of casualties sustained by the Taliban in Lone Survivor was also disputed, but Hollywood always lived by their own poetic license and MC Kerr subscribed to no one's poetic license. Books by Richard Marcinko in 1997 and 1999 accurately represented the SEAL community according to Kerr. Master Chief Kerr was not a fan of former Navy SEALs who promoted themselves through embellished memoirs and exaggerated mission tales. Books by Mark Owen (Matt Bissonnette-real name) and Robert O'Neill that detailed aspects of the SEAL Team Six mission to kill Osama Bin Laden were not high on Evan Kerr's reading lists.

MC Kerr never felt that Deklan Novak had any intention of promoting himself financially by playing college football. PO1 Deklan Novak wanted to play Division I football and was willing to pay a very high price to conquer an admirable challenge. MC Kerr did not believe for one minute that Deklan Novak had attempted to or placed the SEAL community in a bad light or an exploitive light. In fact, the argument that MC Evan Kerr brought to Coronado on behalf of Deklan Novak was that Novak's athletic aspirations were good for the SEAL Teams and the future recruitment efforts to find the right candidates to attend SEAL training or BUD/S. Deklan

CHAPTER SIX

Novak, MC Kerr explained to the SOCOM hierarchy at Naval Amphibious Base Coronado, exemplified everything the SEAL community stood for. Novak was a full-time Navy SEAL instructor at Great Lakes Recruit Training Command, was a husband with three young boys, was a full-time student at Northwestern University and was playing Division I football in a Power Five conference. The sheer dedication to craft necessary to accomplish the schedule laid out was beyond admirable, it was heroic. The SOCOM decision came down to allow Deklan Novak to play college football with no more front-page articles in the newspapers. The one caveat to the football offer was that the job at Great Lakes could not be compromised or neglected in any way. MC Evan Kerr, the Commanding Officer at Great Lakes Recruit Training Command, Head Coach Lance Adams, and Deklan Novak were informed. Novak was back at summer camp three days later.

The NCAA rules for college football eligibility gave students a five-year window to play four seasons. Deklan Novak was entering season one and had no plans to red-shirt for any games or red-shirt for an entire season. The idea had been floated by the coaching staff as an option available to help Deklan ease back into football and hopefully, get the Navy off everyone's back once and for all. In Deklan Novak's mind, no way. At 32 years old, the window to play college football was not opening any wider as the years rolled on. Deklan Novak respectfully declined any option to red shirt for a season, a game, or a day.

"The roster is based on talent." Head Coach Adams reminded Deklan Novak. "You will have a roster spot but playing time is another animal."

"If I can't hold up my end in practice, then you will not have to worry about me hanging around the team to provide emotional support and wisdom. I came here to play ball, coach and if I cannot earn any time on the field then my duffel bag will pack itself." Deklan Novak responded. Novak was at Northwestern to play football. The degree would be great, but the active-duty Navy SEAL was on board to suit up and hit people.

"That is entirely what I expected you to say." Lance Adams replied, stood up, said nothing more and ended the meeting.

There was a buzz at summer camp each year that revolved around the upcoming season. Head Coach Lance Adams had one goal each year and that one goal was pounded into every player that wore a Wildcat's uniform. The goal each summer was to win the Big Ten title. Everything else after winning the Big Ten title had a progression all its own. Bowl games, playoff games and even a national championship were not to be addressed until the Big Ten title was secured. The buzz was greater than usual this year, due to a 32-year-old defensive end/linebacker named Deklan Novak. During the first full-team meeting at summer camp, Lance Adams asked the team if Deklan Novak could carry the American flag out of the tunnel for the first home game. The team went berserk! The entire Northwestern football team, covered in sweat and dirt, tackled the Navy SEAL on their roster before he could take three steps towards the locker room. Beast mode took on an additional connotation in Kenosha at the University of Wisconsin-Parkside.

Before the summer sessions were over, Deklan had solidified a position on the kick-off and kick return teams. Deklan Novak would not see significant playing time as a defensive starter for the Northwestern Wildcats of the Big Ten, arguably, the toughest football conference in the nation. The SEC faithful would vehemently claim the mantle of NCAA deity. Most walk-ons would have considered a role on special teams as a major accomplishment after sitting out football for the past 14 to 15 years. NU was not a ranked team in the preseason college football polls, but Head Coach Lance Adams had put NU back on the football map. Northwestern University was no longer an assumed win on the Big Ten schedules. NU was going to hit you in the mouth hard and the rest of the league had begun to take notice. Deklan Novak's goal was not to simply play a role on special teams, but that was a decent beginning for a ten-day summer camp.

At the end of the first week of summer camp practices, downtime was at a premium, and the players were universally spent from the drills amid the relentless August heat. The team was called to a circle surrounding Head Coach Lance Adams to close the first week. All the players, assistant coaches, trainers, staff members and a few invited guests all gathered around Coach Adams. Everyone took a knee. The heat was unrelenting. Coach Adams surveyed his team and spoke.

"I noticed at seven-on-seven drills that many of you looked gassed." Head Coach Adams began. "I certainly can understand getting tired in this heat, after a long summer break from football and the acclamation to this kind of training. Fuck tired, gentlemen. Do you think Ohio State or Michigan is going to be tired after a 12-play drive on national television in the fourth quarter? Fuck tired. This is where you address attitude. If you bring a weak attitude to my team, you will not make it here. If you bring a weak attitude to the field, you will never beat the man across the line in front of you. Practice builds attitude. We practice plays and situational football. You need to practice your attitude. I don't give a fuck what you say to yourself…just get your head around pushing yourself to a limit that you never knew. Ask Deklan Novak about fucking pushing yourself beyond limits. When you get to the limit you never knew, push fucking harder. We will not get beaten physically by any team we face, or I will put the entire fucking second team or the scout team on the field." Lance Adams paused for a short time. He walked around the large group frozen on one knee with eyes fixed on their head coach. Lance drew a small drink from the water bottle near him on the ground and continued. The head coaches' tone was six notches beyond direct.

"When you step on a practice field or a gameday field, you either get better as a player and a person or you are wasting your time. There are no factors that decide a game outside of yourself. The weather is never too hot or too cold. Injuries don't mean shit. Next man up. The referees never steal a game. The crowd on the road is never too loud. Winning is a choice. Pat Tillman maybe said it best… **Somewhere inside, we hear a voice. It leads us in the direction of the person we wish to become. But it is up to us whether or not to follow**. I'll see you after lunch today and we'll be back out here tomorrow morning. This was a warm-up, gentlemen. The hard work has yet to arrive."

Coach Adams did not wait for anyone to approach him. The team began to move slowly and cautiously. The head coach was off on a rapid paced jaunt to the main buildings set up for the coaching staff next to the gymnasium. Lance enjoyed walking alone. Adams never wanted anyone patting him on the back to tell him that he ran a good practice today or that his talk after practice was terrific. Teams practiced improving. Teams listened to learn. There was nothing else necessary. Football players had become too full of themselves, and the adulation directed their way was a distraction. Although Lance understood the haughty connection between success and publicity, strong men believed in the team first and always. Lance believed in few clichés, but he knew weak men played on weak teams and weak teams lost.

The practices that followed did not disappoint. The driving physicality of the drills in the heat was relentless. The team cherished the evening meals and the short time they had to themselves in their bunk units. As each day passed, Deklan's Unit #7 and the dormitory floor lounge next to Unit #7 became the gathering place for most of the defensive team. The players bonded through their efforts on the field, and they bonded around the Navy SEAL whom they coveted to know. Deklan was never an outgoing social blatherskite. Deklan was not one to relish sharing his deployment missions with strangers or family for that matter. Eventually and after the questions had become relentless and redundant, the Big Ten rookie hesitantly spoke of a 2010 Special Operations SEAL Team One mission in Afghanistan that had some relevance to his new teammates.

"The assault team target was an insurgent mountain stronghold between the Safed Koh Mountain range in eastern Afghanistan and Jalalabad. Bin Laden's ultimate demise occurred a few years later at a compound just outside of Jalalabad." Novak never felt comfortable with everyone's undivided attention. The recantation would be condensed and as generic as Deklan Novak could articulate. Combat details seemed to always be subjected to judgement from non-participating, self-appointed pundits.

"Our target was in Pachir Aw Agam province. There was a dozen SEAL Team One guys and 50 Marines out of the 15th Marine Expeditionary Unit (MEU) who followed us in armored Humvees equipped with SPG9 and DShk, anti-tank weapons and heavy machine guns. I stood in awe at the base of the Spin Ghar mountain range silhouette. Even at night, the cold bleak crevices held the blood from many battles. The carved rocky outcrops lay in a great line like the backbone of a forbidden zone. The location had become a staging area for Taliban fighters and other insurgents preparing attacks on NATO and Afghan forces. Some 300 Taliban had assaulted the American Combat Outpost Keating some weeks before we arrived. The Taliban attack left 8 Americans dead, and more than two dozen wounded. The outpost was subsequently evacuated and bombed to prevent looting of the munitions left. The Taliban exaggerated the success of the attack on their website, Voice of Jihad as usual. Our assault team fast roped into the area after midnight. Wearing night-vision goggles, we made our way up the terrain and near the sleeping unsuspecting insurgents. The Marine Humvees were somewhere below us, but my job was not connected to the Marine convoy progress. As a breacher, my job was to open whatever doors we happened to find between our platoon and the bad guys. That might have been a hidden complex on the mountain or a fortified bunker inside a cave. We had an AC-130 hovering above outside of earshot range for support if we needed the air support. The AC-130 was a flying artillery version of Amazon.com. The assault team could punch in what was needed and the package was delivered in short order. The AC-130Us had a pressurized cabin, allowing them to operate 5,000 feet higher than other models, which resulted in greater range. These flying gunships could shoot off the top of a mountain. All our actions were to protect our brothers and the United States of America." Deklan paused and scanned the room. People eventually asked enough times about combat but returning to those events remained difficult for Novak. Death was not a circus to place under the big tent.

"Expectantly, we observed movement from the hillsides." Deklan continued. "Something spooked our targets from their fragile sleep, and we could see them arming themselves with AK-47's and RPG's. Wild gunfire was sent our way from the undisciplined forces so unceremoniously awakened. I watched as one of our Afghan guides was hit directly with an RPG. RPG's generally travel at 1,000 feet per second or the same speed as a .45 caliber pistol round. Most RPGs are constructed to detonate when a hard target is hit like a tank or a Humvee. When an RPG strikes a human being, the round may not detonate immediately because the target was not a hard target. If the human target was wearing body armor or the round struck the helmet, detonation occurred immediately. Time delay fuses inside most RPG's triggered detonation if the round missed an intended target and came to rest intact. Regardless of when the shell was going to detonate, our Afghan guide was obliterated by the RPG, and he was less than twelve feet from where I was crouched. We had to act." Deklan perused the room again and looked through the faces before him.

For a moment at the University of Wisconsin-Parkside, Deklan Novak remembered going back to the recruiting office where he signed up to become a Navy SEAL. "Why do you want to be a Navy SEAL?" the recruiter asked Deklan on that day. It probably didn't matter what he said, but Deklan remembered what he told the recruiter. Deklan told the recruiter that he wanted, more than anything, to become part of something elite, something unattainable. Deklan Novak remembered again what Muhammad Ali said: **Impossible is not a fact. Impossible is an opinion.**

"The red laser dots began to find their targets and the SEAL snipers began to fire round after round." Deklan resumed the narrative. "I watched my SEAL sniper teammates in awe. The return fire was erratic and faded as the target group fell man after man. As we got closer, the assault team snipers continued firing their McMillan TAC .338 caliber sniper rifles at movement within the compound. The 5 round magazines obliterated the intended targets. The common effective range for a TAC .338 is 1600 meters but unique talents have extended that range. The TAC .338 was the rifle used by Chris Kyle when he killed an insurgent at 2100 meters. Navy SEALs do not shoot astray. Navy SEAL snipers allowed Novak's unit to slowly gain ground on the compound and they reached the target area once the movement and the return fire had stopped. The scene was horrific. Navy SEAL snipers were the most accurate shooters in the world. Navy SEAL Sniper School was three intense months of 12-hour days, seven days a week training that followed BUD/S and was arguably more stressful than combat. Novak viewed the carnage on the grounds. Men had been blown to pieces by the sniper rounds. Skulls were torn open as they stepped over the scattered bodies and the rugged terrain was covered in a sea of blood. The assault targeted insurgents would not be attacking anyone else." Deklan paused. The room was crowded around him and silent.

"The insanity of SEAL training in Coronado was to prepare Navy SEALs to be the most physically superior Special Forces team in the world. There is no possible chance to survive BUD/S training without the right mental attitude. BUD/S was only the start to training as a SEAL. My SEAL Team One commander worked us like BUD/S was the freshman team tryouts. You're a SEAL now asshole, we train every fucking day we are not on a specific mission. While the brutal physical challenges of SEAL training produced superior athletes, the training also prepared each man for what they were going to see daily during combat deployments. The physical strength gained through unimaginable sacrifice was the psychological foundation unconsciously forged when we stepped over a body that was completely unrecognizable as a human being. War is messy. Confusion runs rampant. Battle grounds get sloppy, and we lose many soldiers and allies to friendly fire. Pat Tillman died by friendly fire when his "Black Sheep" platoon was split up after a GMV (Ground Mobility Vehicle) stalled on an isolated road in a mountain valley not unlike the location we had found ourselves in that night. Fear, disorientation, and chaos existed under fire when soldiers were screaming over volleys of Rocket Propelled Grenades and 14.5 mm Taliban machine gun fire to call in the co-ordinates for air support. Mistakes were made often when positions were compromised, and Taliban forces were firing on various coalition platoons from invisible perches in the mountains. In June and July 2010 there were 5,500 close air support sorties in Afghanistan alone flown over ground troops locked in active combat.

On 900 of those flights, the planes fired weapons. Drone strikes were a desirable alternative to U.S. military ground raids into the jihadist protected terrain. As effective and necessary as the drone strikes were and the air support strikes were, it was impossible to determine the number of friendly fatalities and non-combatant civilians that were killed by mistake. Training is all about eliminating mistakes. Pushing beyond your own limits is what challenge is all about. How far you can push will ultimately define what you can accomplish. Coach Adams was correct. Different level. Same premise." Deklan finished and did not focus on anyone in the room directly.

Deklan was done for the day recounting the activities that he normally was reluctant to share. The acclimation to the football team was rewarding in unexpected ways for Deklan Novak. The younger teammates found something they weren't looking for, as did Deklan. The players began to break up for the night. No one pushed the older linebacker for more than he was willing to give.

CHAPTER SEVEN

2013...DEFENSIVE COORDINATOR BEHR THOMAS

THE UNIVERSITY of Wisconsin-Parkside campus was quiet at night in the middle of August. The Northwestern University football team were the occupants of the athletic complex for eleven days. The 10:30 p.m. lights out summer camp edict was not a prison blackout at bedtime. The televisions ran for some time after 10:30 and the video game consoles buzzed in the dark dorm rooms as players snacked and dreaded the next long day to come. Deklan Novak wasn't ready to go to sleep and neither were two other linebackers glued to Assassin's Creed IV on the television screen in his room. Novak walked down to the main lobby and onto the courtyard outside of the main gymnasium and makeshift locker room for the football team.

The courtyard outside of the main gym entrance was a large circular complex of slate flat rocks surrounded by a small two-foot-high slate wall of slab rocks that acted as a half-moon bench for the courtyard. The center of the courtyard was highlighted by three large flagpoles rising over twenty feet in the air. The American flag was in the center and was flanked on both sides by the state flag of Wisconsin and the flag representing the University of Wisconsin. The air was thick with humidity and the summer night had done little to lower the temperature. Defensive coach Behr Thomas was sitting on the slate rock wall near the far end of the courtyard and was obviously on the phone. The large parking lot facing the athletic complex was mostly empty except for the coaches' vehicles. The players all took the bus from Evanston for the start of summer camp. The dorm room accommodations had been brought up during a two-day move in window, but the team left for camp together from the NU campus. Players were not allowed to leave the complex during the summer training camp. Prohibiting cars was always a deterrent to nighttime unauthorized excursions.

Deklan sat at the other end of the courtyard. It was 10:45 p.m. The moon served as a partially clouded bonfire to highlight the surrounding Balsam Fir trees and the Bigtooth Aspen trees that guarded the facility. The trees rose to almost 70' above the UW Parkside campus buildings and the courtyard crackled with each step as Deklan hit the edible, brown hard beechnuts that had fallen from the American Beech trees nearby. Deklan wore his UDT Navy shorts, no shirt, and flip-flops. He sat down to listen to the message from his wife again. Amber was threatening to take the kids back to her parents in Kenilworth. The fallout from another reality dose of Deklan Novak's isolation from the family had to be addressed and Amber had reached her breaking point. Deklan listened to the message a couple times while the night air and his anger caused the sweat to roll from his forehead onto his chest. Deklan put the phone down and stared straight ahead. The phone conversation from his defensive coordinator was impossible to ignore.

The former University of Michigan safety and former linebacker for the Chicago Bears was on the phone with his ex-wife, Raya Nolan Thomas. Behr was nearly the same age as his walk-on linebacker from the United States Naval Special Operations Command. Coach Thomas was not concerned with Deklan Novak sitting across the courtyard from him.

"What do mean, Raya, god dammit." Thomas asked with managed restrained anger in his voice. "Kaley has been in the hospital more often since you left than my whole team combined.

CHAPTER SEVEN

What you are telling me now is that Kaley had some kind of abdominal blockage that the doctors cannot determine what caused it again. This is the fourth time this year, you had to take Kaley to the emergency room. Why don't you bring her back to Evanston. The staff at Northwestern Memorial Hospital will fix her for sure. I'm tired of having the same conversation with you. Kaley has not been doing well since you guys left. Bring her back and we can get her the right care." Behr finished and wanted to say so much more, but Behr Thomas was not calling the shots with his daughter. Deklan sat silently and listened to Behr's anguish, not from an eavesdropping curiosity but from an empathetic father, unfortunately about to run a similar course. Behr looked up at Deklan but said nothing. Behr was listening to his ex-wife and the conversation did not appear to be improving.

"Raya, you cannot argue that since you arrived back in Steamboat and hooked up with Gallardo, Kaley has done well." Behr paused.

"No, I'm not saying it's your boyfriend's fault. Listen to me. The environment has not been good for Kaley. I don't know why, but just bring her back here for some medical tests and at least then, we can know what has been going on." Behr waited again.

"Carmen is not helping you down there." Behr raised his voice. "The guy is an alcoholic and doesn't have a full-time job. How is that a good place for Kaley?"

"God dammit Raya." Behr shot back. "Don't go over that again. We moved around. Big deal. Coaches move around until they land their dream job and it's all worth it, but you could never understand the concept of…" The call was cut off.

"Shit!" Behr Thomas yelled at the dead cell phone. "Mother fucker!"

Thomas looked up at Deklan again and the two men stared back at each other for a few odd silent seconds. Behr Thomas was not about to scold his player for breaking the lights out arbitrary rule guidelines. Deklan was not about to insert himself into another marital calamity. One was his limit. Behr broke the silence.

"It's still fucking hot here this late. Let me guess, your roommates are not sound asleep, and X-Box is not your thing?" Coach Thomas asked and didn't care.

"Pretty much true, coach." Deklan replied. "I came down to try and call my wife. She left me a voicemail that wasn't what I had been hoping to hear after a couple days of summer camp. She didn't ask me how it was going or if the training was anything like SEAL training. She basically told me that she would be leaving me soon and taking our boys back to her parents."

'I've been divorced for over a year." Thomas confided. "My wife took my daughter back to Colorado where she is from. My daughter is six years old, and she has been sick on and off since they moved back to Steamboat Springs."

"Sorry. Coach." Deklan answered without digging. If it was to come, it would come.

"My daughter's name is Kaley, and she takes my breath away when I look at her. She's beautiful and I still cannot believe I had something to do with making something so breathtaking." Behr Thomas paused to gather himself professionally. "My ex-wife, Raya, has had to take Kaley to the ER on more than one occasion since they moved back to Steamboat. Kaley has had a tough time keeping food down and she has been throwing up constantly, where the ER visits have become necessary. They treat her, cannot pinpoint what has been going on except to say that she has some form of gastroparesis, whatever that is? Apparently, gastroparesis is a disorder that prevents the food in the stomach from passing into the intestines and eventually out of the body. The condition can produce vomiting, nausea, and severe abdominal pain. I have researched this a ton, and the experts cannot say for certain what causes the condition. They treated her at the hospital, and she seemed to improve quickly, but the condition has reoccurred again and again."

"Is she getting good care in Steamboat Springs?" Deklan asked as he moved closer to his defensive coordinator.

"I guess." Behr answered grudgingly. "My insurance is great, so there have been no issues with coverage. Raya has legal custody, so I am not asked to consult on medical decisions. My ex moved in with an old boyfriend and he is an alcoholic shit bucket, but that's also none of my business. How long have you been married?"

"Amber and I got married when I was in BUD/S training back in 2006, I think." Deklan honestly did not remember his anniversary date. "Amber was there for the physical nightmare in BUD/S and then after I was assigned to SEAL Team One in San Diego, she stayed there during my multiple deployments and training assignments with SEAL Team One. I was gone for probably 280 days a year. We moved back to the Chicago area with my assignment to Great Lakes. My decision to enroll at Northwestern and play college football has not been a popular decision at home or with anyone in the family from my mom to her parents to my uncles, siblings, and grandparents. They think I'm trying to prove something. I'm not sure if that is a bad thing? Aren't we all trying to prove something every day? I don't have those answers, but I imagine the resentment may come from one's own failures to pursue their own dreams and aspirations. I try to live my life, so my boys have a blueprint worth following."

"How does your script end?" Behr asked.

"That's a good question, coach." Deklan pondered the inquiry briefly. "To be determined, I guess. How does your script end, coach?"

"To be determined as well, I guess." Behr replied. "My script has a short window moving forward. That, I know for certain. Legally, I must start making some plan to get my daughter back to me. That's not going to be easy or cheap, but it will get done. I'm going back up to get some sleep. You should follow me. Practice doesn't get any easier going forward. You are making good progress Deklan. Get some sleep PO1 Deklan Novak." Behr stood and nodded that short, abbreviated, upward head toss that guys give to other dudes as a sign of friendship or approval. Behr wanted to salute Deklan, but Behr had already researched a civilian salute, which was considered disrespectful to a military person. A salute is a gesture of respect that is reserved for military personnel as a way of acknowledging rank and authority. The two men headed back to the complex. The beechnuts crackled again under the shuffling feet amid a sprawling bed of lustrous dark green Balsam needles.

CHAPTER EIGHT

2006 NAVY BOOT CAMP AND A ROAD TO NAVAL AMPHIBIOUS BASE CORONADO (NAB CORONADO)

DEKLAN NOVAK HAD NOT LOST his focus from the last trip to the Navy recruiting office. There had been a stint in college, a construction detour and a successful beginning to an MMA journey that eventually may have brought Deklan Novak to the UFC, Dana White's Greatest Show on Earth, and the biggest stage in mixed martial arts. After Deklan's fourth victory in MMA in Cicero, the impetus changed for the up-and-coming new fighter. Deklan's new confidence did not lead him to start dreaming of championship belts and fighting in the UFC main events on Pay-Per-View, ESPN, Fox, or Ultimate Fighter telecasts. The victories in MMA proved to Deklan in his mind that he was now tough enough to become a Navy Seal. How was Deklan going to make the biggest impact on the world and how could he best make a tangible difference to what must be done. Deklan Novak knew he had to become a Navy SEAL. They were the best, the most elite fighting organization in the world and wearing a SEAL Trident meant a lifetime of respect. The MMA training transitioned into SEAL training. Weights and combat wrestling became running in the sand with weight belts, swimming, endless push-ups, and endless sit-ups. The MMA became secondary to joining the hardest club in the world to join.

The Navy contract had to be in writing. Deklan was a young, dumb kid, and he was signing five years minimum of his life away based on a recruiter's promise to give Deklan a shot to go to BUD/S after boot camp. The deal had to be in a contract, signed sealed and unbreakable. Finally, Deklan had joined the United States Navy. Deklan Novak was going to be in the first BUD/S class, Class 264 that would be allowed to go from Great Lakes directly to BUD/S in Coronado without going to A-School first, usually a three-month job training selection necessary because most all BUD/S candidates failed to make it through the training. There was a war on terror growing and the SEAL pipeline was thin, therefore, the Navy decided to make sure there were always qualified candidates in the SEAL system. Deklan was headed to Recruit Training Command at Great Lakes.

Deklan Novak arrived at Great Lakes Recruit Training Command or Navy boot camp in late summer 2006. All the new Navy recruits were picked up by a Navy transport van at O'Hare International Airport. Driving up to the Great Lakes complex as the sun rose high above Lake Michigan to the east, the red brick sign structure left no doubt where the van had arrived. *Naval Station, Great Lakes, Illinois, Recruit Training Command, Camp Porter* announced in block white letters that the caravan had entered a United States military compound. The moment the van passed the main gate and pulled up to Golden 13, the building where recruits began their careers in the Navy, Deklan wondered if he had made the right call.

P-days or processing days for The United States Navy began with the usual expectation. Recruits were examined in their street clothes once they entered Golden 13 for the first time. Naval officers prepared the new recruits for the days to come. Recruits were examined for offensive tattoos. In the old days, a recruit was not allowed to enter the Navy with a tattoo that extended lower than the elbow. Those days were gone. Tattoos, however, could not be offensive in any manner and many Navy recruiters missed those tattoos. Tattoos could not represent gang affiliations. Tattoos could not represent any

CHAPTER EIGHT

white supremacy movement or extreme nationalist movement. Tattoos could not show Adolf Hitler's birthday as in the 4/20 numbers. Tattoos could not show the numbers 311 representing the 11th letter in the alphabet "K" multiplied by 3 or KKK. The 100% tattoo was banned from the military representing that the individual was 100% Caucasian.

The first day at boot camp for a recruit involved exorcizing from each recruit, their ties to the civilian world. Each recruit gave up everything brought to boot camp. Cell phones, gaming devices, hats, clothing, jewelry, and any writing materials. The recruits kept only their wallets. All their belongings were boxed up and shipped back to their parents or other relatives. The recruits were allowed a two-minute phone call from a pay phone telling their loved ones that they arrived at boot camp. Deklan called no one on the day he arrived at Recruit Training. The next call would come in three to four weeks. Medical visits followed as did a mental evaluation that gave the recruit one final chance to admit if he or she had lied to their local recruiter. If any recruit had a police record uncovered following their interaction at their local naval recruiting office, the recruit was dismissed and sent home. If the recruit lied about anything on their original application, they were dismissed and sent home. Haircuts followed the MEA culpa interviews. Boot camp clothing issue followed the head-shaving assembly line, and the procession eventually marched outside with Recruit Division Commanders barking out the orders that the new recruits would become very accustomed to hearing during the next eight weeks.

All the recruit barracks buildings were named after Navy ships. Deklan was assigned to the USS Triton. Recruit Training Divisions had anywhere from 50 to 90 recruits. Men and women were part of the same Divisions. Deklan was assigned to the 900 Division, a division made up of the highest scoring ASVAB recruits. ASVAB is the Armed Services Vocational Aptitude Battery, a proven method of determining who is a good fit for the military and where they will be a good fit. The divisions each had three Recruit Division Commanders, and the division chose recruits to fill the following positions:

Recruit Chief Petty Officer, Recruit Leading Petty Officer, Recruit Master at Arms, etc., etc. Recruiting candidate assignments to division positions of authority was the first step to teaching the new Naval candidates that they were part of a team from that day forward.

Modern boot camp regulations for the military branches and specifically Navy boot camp had been forced to comply with guidelines handed down from Washington. Soft guidelines were drafted mainly by men who had never served in the military. The main restrictions now placed on boot camp military training began with touching a recruit. All physical contact with a recruit was banned from boot camps in the United States. Boot camp instructors or RDCs in the Navy were additionally not permitted to assign supplementary physical exercise as a means of individual punishment. Profanity was banned from use in boot camp when it was directed at a single individual. Instructors could swear at the entire division but not at one recruit. Finally, recruits were told that once they signed their Navy contracts, they were not allowed to leave the Navy. That was incorrect. The statement was true after boot camp. A recruit could quit boot camp at any time by claiming the stress was too much to handle and he or she would eventually harm themselves if not allowed to leave. Many recruits chose the path with no intention of ever harming themselves. The extraction process took about a month to complete, involved a cold, heartless separation, and humiliation from the assigned division, but there was an out for a recruit after he or she signed their Navy contract.

Deklan had a tough first few days at Navy boot camp. The experience was a complete culture shock to Deklan. A recruit ceased being a person. Recruits were numbers and were marched around like robots without souls. Deklan Novak knew he had made a mistake and less than one week into boot camp, Deklan walked into his RDC's office after the lights went out on night five and was crying.

"You gotta let me get out of here, sir." Deklan begged with tears in his eyes. "This is not for me. I made a huge mistake, and this has got to stop now. Please, I need to go home, sir."

"Deklan, relax." The RDC stepped out of his screaming persona

CHAPTER EIGHT

and made sure the two men were alone. "Give it a few more days and things will calm down. If you feel the same way in 3-4 days, come back and we can talk again. Now, I know I have a special recruit in you, Novak. The scores you put up, the resume you brought here, and the SEAL contracts are not handed out like candy. What the fuck am I going to look like If I can't get a Navy SEAL candidate through the first week of boot camp? Give me a few more days, Novak. Can you do that?"

"Yes, sir. I can" Novak responded and inexplicably felt better.

On the walk back to his bunk, Deklan scanned the barracks and prayed everyone was asleep. How the fuck could an MMA undefeated fighter walk into his Recruit Division Commander's office, crying and ask if he could go home? Deklan wondered if he should look for a dress to put on for the next day's training schedule. Deklan decided that his MMA skills were possibly going to be called upon if there was anyone awake in the 900 Division that witnessed any tears falling or begging in progress. Those skills were not called upon that night.

Two days later, Deklan Novak had an epiphany that would define his life. Once Deklan accepted the reality of where he was, he loved it. The culture shock of boot camp was the loss of self. The military had to break down every recruit before they instill the mantra of team first. Once the yelling and demeaning boot camp drill instructors were not taken personally, the mantra became a competition to teach the screaming drill instructors that they were not scaring anyone. The team prevailed every time and when the division got it, the camaraderie was forever. Boot camp was candidates competing against each other all day long. Deklan loved competition. There was plenty of food available. Everyone was able to work out all the time. There were girls in the division. The 900 Division was home to the Special Warfare Candidates or SEAL contract candidates. The Special Warfare candidates had to wake up two hours before the division to train at the pool for SEAL training exercises, then they began the normal early boot camp regimen. Deklan was now getting paid to train. Life made sense.

Week seven at recruit training was Battlestations 21. Battlestations

21 was located on the USS Trayer (BST-21). Building 7260 or the USS Trayer was a 2/3 scale, 210-foot mockup of an Arliegh Burke-class destroyer designed by the most talented set designers in Hollywood complete with Hollywood special effects. Recruits were required to endure a 24-hour test to complete recruit training that included 17 real-life test case scenarios. The purpose of the grueling final week test at recruit training was to put the recruits in situations that changed within a heartbeat. Recruits were tested when the ship had a simulated bomb slam into the main hull sending torrents of water and shrapnel into the ship's barracks. (recreated in response to the 2000 attack on the USS Cole) Recruits had to react to the ship filling with smoke at a moment's notice. Recruits at the end of week seven discovered where their skills stood inside live fire drills, flooded ship scenarios, counter-terrorism attacks at sea, deadly gas attacks, electrical power compromised situations and many other life-threatening potential circumstances that occurred at sea.

The end of Battlestations 21 led the participating recruit Divisions including Division 900 and Deklan Novak to the cap ceremony where they could discard their Recruit caps, and they were handed official Navy caps. The recruits had made it through recruit training and were now sailors and part of the United States Navy. The ceremony was always an emotional release from the seven weeks of misery and the stress of Battlestations 21. Many recruits had been diagnosed with PTSD from participating in Battlestations 21. The cap ceremony culminated with speeches from the Commanding Officer at Great Lakes and other visiting dignitaries standing at a podium framed by the American flag and the Navy flag. Immediately following the speeches, the crystal-clear public-address system thundered out the resounding Navy anthem heard at the conclusion of every Recruit Training cap ceremony. Lee Greenwood's "God Bless the U.S.A," captured the room and filled the newest members of the nation's armed forces with humility and pride. Tears flowed like rushing rivers in the room. Recruits hugged and embraced each other. Goosebumps stood at attention on the arms of each recruit. Men and women forgot about the seven weeks they endured, and the room became an entan-

CHAPTER EIGHT

gled sea of sweat soaked, smelly, delirious members of the United States Navy. Deklan Novak wiped the tears and the dirt from his cheeks. He took off his Navy cap, curled the bill slightly and placed it proudly back on his head.

Boot camp graduation was a grand traditional ceremony every eight weeks at Great Lakes Recruit Training Command. Deklan's parents, his brother and sister, his grandparents and many other relatives attended the boot camp graduation ceremonies. The divisions were introduced, awards for physical fitness and meritorious honors associated with drills, marching, and discipline were handed out. Parents beamed and the pride poured from the auditorium like liquor and glitter in New Orleans on Mardi Gras. Amber was there and watched the man she loved enter a world that would change them both in the most unexpected ways.

There was no Pre-BUD/S class at Great Lakes that helped the new boot camp graduates and Navy SEAL candidates transition to the Strand in Coronado. Pre-Buds became a two-month class that Deklan Novak eventually taught after nine years on SEAL Team One and while he was a full-time Northwestern University student and a member of the football team. After Deklan's graduation from boot camp, there was a three-month limbo time frame until BUD/S Class 264 was to begin. Deklan and six other SEAL candidates from the boot camp class stayed at Great Lakes and basically trained on their own to prepare for BUD/S. The SEAL candidates became obsessed with arriving in San Diego ready to take down SEAL training without a hitch. Deklan and his buddies were clueless, but they trained like they knew the drill. The crew worked out all day long. They went down to the beach at Lake Michigan and sat in the cold water to prepare for Hell Week. They brought small boats down to the beach and practiced rock portage, where they carried boats overhead while navigating rocky shores. The hearty crew of six practiced getting wet and sandy, the orders at BUD/S to run down to the ocean, jump into the ocean and then roll around in the sand until every pore of your body was covered with sand including your head, your genitals, and the inside of your ears.

The crew of six became tight friends and pledged to stay together through BUD/S training and they vowed to work together on the SEAL Team assignments that were certain to come down the road. The reality was that Deklan, and one other SEAL candidate were the only ones to survive BUD/S training. The four other boot camp compadres rang the bell at Coronado before the first week had ended. Deklan Novak and John Cook left for San Diego on an epic journey that would eventually culminate at the hardest training regimen in the military.

CHAPTER NINE

2006-2007 UNITED STATES NAVY'S SEA, AIR, LAND TEAMS (SEALS) AND THE SELECTION BY NAVAL SPECIAL WARFARE COMMAND.

DEKLAN NOVAK ARRIVED on the Silver Strand in Coronado to begin BUD/S training. (The rate of advancement from contacting a recruiter through the enlistment process, ASVAB testing, recruit training (boot camp), completed Pre-BUD/S INDOC, completed BUD/S Phase 1, 2, & 3, Airborne School completed and completed SEAL Qualification Training (SQT) is approximately 71 weeks of training and 9.5% of the total number of candidates who signed up to become Navy SEALS will make it through to receive their Trident. Simply put, for every 100 men who attempt to become Navy SEALS, less than ten will succeed.)

BUD/S: THE FIRST DAY ON CORONADO ISLAND.

John Cook and Deklan Novak drove onto Coronado Island just before 2100 (9 p.m.) on Sunday night for their Monday morning start to BUD/S. They were reckless and excited. Both men desperately wanted to become Navy SEALs. Both men felt they were tough enough and had what it would take to survive BUD/S and beat Hell

Week. The two men were also young, wild, and full of adventure, so rather than check into Naval Amphibious Base (NAB) Coronado an hour earlier than required, which was already cutting things close, the pair of SEAL candidates crossed over the San Diego Bay on the beautiful arching Coronado Bridge for a mandatory island tour before signing away their freedom.

John Cook and Deklan Novak walked through the doors to the United States Navy SEAL Basic Underwater Demolition School at 2158, two minutes early. This was the location where they hoped to become men. Inside the complex were pictures and memorabilia of all the fallen Navy SEALs. Tridents and frogmen adorned the walls. The equipment and weaponry and evolutions of the Teams were displayed everywhere. Murals of BUD/S candidates engaged in every kind of horrific evolution hung from every corner. They moved with baby motions and wide eyes as the excitement from the room was only overmatched by the fear it represented. Deklan had another thought. Where was everyone?

The sound of a toilet flushed from behind a corner was heard. John and Deklan froze. A door was opened and the two SEAL candidates froze like statues in their tracks. Were they in trouble? Were they too late? Were they about to be tossed before they started the program? Maybe they would have been better off arriving early? Was there still time to just back out of the door and call it a day or choose another job in the Navy, an easier one? Footsteps approached and Deklan prayed that the man approaching was not the infamous instructor, Drago. To their surprise, the prettiest, cutest female Petty Officer stepped out of the back hallway in her well pressed working camouflage fatigues, sparkling black boots while drying her hands on her pants.

"Are you here to check in?" She asked and smiled.

"Yes, we are." Novak replied. "Seaman Recruit Deklan Novak and Seaman Recruit John Cook reporting for duty."

The female Petty Officer looked down at her list and confirmed the names were on the list. John and Deklan were the last two to check in. BUD/S candidates 273 and 274 of Class 264. Class 264 was the largest BUD/S class of all time. That was it. They were given instructions to the barracks where rooms were assigned, and gear was handed out.

It was like Lord of the Flies. It was like a Romanian Prison Camp. It was like an SAE Frat house. It was like a beachside men's commune. At 2200 on a Sunday night, everything was stirring. The only thing between Novak and the Pacific Ocean was a 15-foot sand dune and 100 feet of beach. Even the ocean was stirring. With the sound of 8-foot waves crashing onto the beach, the music played from a myriad of rooms that were full of men scurrying around with brooms and mops and shaking out rugs. They were folding clothes and obsessively wiping down every inch of flat surfaces in the building. There must have been an inspection coming on the first morning. Deklan asked a kid in a white t-shirt where he could find the Leading Petty Officer (LPO). The kid pointed towards the washing machines and a glass door.

Most SEAL candidates had planned for years to become a United States Navy SEAL. The commitment necessary to join the most elite military unit in the world did not usually occur based on a sudden epiphany. Most Navy recruits and SEAL candidates heard their aspirations of becoming a Navy SEAL trampled on for as long as they had their sights set on Coronado. Teachers, friends, family members and casual acquaintances of each candidate had read about the staggering failure rate associated with SEAL training. The most common response from those skeptics was jealousy. No one was envious of the physical torture candidates incurred in Coronado, but the haste to disparage anyone who thought they could endure the training had always been a defense mechanism to mask envy for the attempt.

Most SEAL candidates chose to sign up for the toughest training on the planet, therefore when the men known as First Phase Instructors or the pain brokers of BUD/S training pushed the SEAL candidates beyond the limits of human tolerance, the candidates fell back on the mindset that they knew what they were signing up for. Indoctrination previewed the physical nightmare ahead. INDOC could not instill the instinct to survive BUD/S. Much like a quarterback can possess all the physical skills to play the position. However, the instinct to sense a linebacker barreling in from the blind side was not something that could be taught. A QB has it or he does not. The physical challenges in SEAL training were perhaps the hardest in the world and the only way to survive physical pain so severe was to push the pain from the mind. Run-times, calisthenic statistics and obstacle course personal records were meaningless without the right mental frame of mind to withstand what was about to occur at BUD/S.

INDOC was the first taste of what BUD/S had to present. In the INDOC phase, Deklan and John Cook and Class 264 were taught how to do the tasks necessary to survive Phase One, Two, and Three. There were techniques to learn on how to swim in the ocean, how to master the boat skills, how to run properly, how to muster (grinder head count line), how to run the obstacle course, what uniforms to wear and how to wear the gear, how to lineup, etc.

Class 264 had the drill instructors licking their lips at the prospect of 274 candidates. Many of them imagined the SEAL Teams were going to be diminished by so many recruits. Others completely enjoyed the ability to get someone to quit. Nothing made them happier, not from spite but from eliminating someone who did not belong. Those eliminations made the Teams safer and more effective. Deklan entered BUD/S knowing he had the constitution to be a good Navy SEAL and Deklan knew he was willing to die before he quit the program. That was not a mantra. That was who Deklan Novak was. If one believed they belonged, then the instructors could not make that candidate quit.

The brass bell was the symbol of failure in BUD/S. All a candidate had to do was to ring the bell three times, and the torture would end. Once the training became too intense to tolerate, the bell was the one-way ticket out of BUD/S. Day one of First Phase was designed to get rid of the entire class in one day. Deklan sensed that the horror stories he had been exposed to surrounding BUD/S training were tame compared to what he would be exposed to before breakfast. Seven candidates in Deklan's SEAL Class 264 rang out on day one before they had reached their first meal as a SEAL candidate.

CHAPTER NINE

The Coronado BUD/S training program was a prescribed 24 weeks of training, divided into three unique phases of instruction. First Phase focused on the physical side of training, combined with shock and awe methods of instruction. Physical conditioning and strength training were the goals for the SEAL candidates while instructors weeded out the weak and uncommitted. Phase Two, Dive Phase, consisted of combat diving skills. Students learned the basics of being a combat diver and were introduced to two types of diving systems. The final phase of BUD/S, Third Phase, also known as land warfare phase, introduced the SEAL trainees to weapons and demolitions while also teaching the basics of navigation and small unit tactics. Although each phase had a unique goal and focus, the trainee was exposed in each phase to common challenges set in place to instill in the SEAL candidate some particularly important attributes that stayed with each candidate throughout his career as a Navy SEAL. Those skill sets included teamwork, loyalty, attention to detail, and self-awareness.

The pre-dawn days began with unrelenting drills and exercises. The candidates lined up on the beach to start calisthenics before the sun had risen. Hundreds of push-ups and hundreds of flutter kicks until the grueling punishment began to feel like everything underneath the skin was on fire. The instructors yelled for the teams to hit the "surf zone" which meant running 500 yards to the 50-degree water and then taking a nose-dive into the sand. The instructors screamed for the teams to return to the grinder position for more unrelenting punishment. The brutality of pushing the body to such extremes was a daily routine. The sand inside every orifice of the body tore up the flesh with every movement that was smothered in a torrent of sweat and dirt. Deklan's body began to break down during the first week of BUD/S. It didn't matter, Deklan was not ringing out.

The next few weeks got worse, much worse. The runs and swims ramped up to more difficult levels every day. The candidates were running close to 20 miles each day and much of that was on soft sand while wearing boots and dragging a ruck sack weighing close to 70 pounds. Deklan was forced to absorb a non-stop barrage of degrading, image bashing, self-esteem wrecking balls thrown at him and the other candidates through each excruciating drill. The candidates never examined how to survive the seven weeks as a long-term process or they rang out in short order. Deklan's thoughts and most candidates zeroed in on surviving the next hour or how to make it until the next meal. Pain was dispersed for pain's sake. SEAL candidates discovered pain and its resistance to language. The physical pain they endured could not be shared as a group. Their physical pain destroyed language. Their pain brought about a reversion to a state devoid of language and to a state the human body found before language was learned. Adjectives no longer defined the pain the SEAL candidates faced.

Deklan discovered something remarkable after a week at Coronado. Deklan enjoyed the challenge. SEAL candidates had weekends off during First Phase training at BUD/S. Most of the candidates sought relief from the pain they had endured for the past five days. Sunday was always spent getting a candidate's gear ready for the coming week and the upcoming Monday morning inspection. Bunks, clothing, and the barracks had to be perfect. Deklan often chose not to venture off the base on weekends with his classmates. Saturdays were spent practicing drills with his boat crews or working on the Obstacle Course times. Deklan and some of his California classmates occasionally explored the San Diego bar scene on the weekends. Surfing was introduced to the Midwestern former MMA fighter on the Strand and Deklan learned to boogie board and bodyboard with John Cook, his California-born brother.

Time became the gauge for survival. Short term time became measured incessantly. Deklan kept track of how much time was left in each drill. Deklan kept track of how much time was left in each day until a meal. SEAL candidates ran almost a mile to each meal. A herd of more than 200 filthy, sweat soaked, sand covered starving maniacs barged into the mess hall three times a day. Pools of sea water gathered beneath each man waiting to be served. Some men fell asleep standing up in the food line. Meals were shoveled down as rapidly as possible to cram down as many nutrients in the allotted time as possible and to steal five or ten-minute naps behind the facility in a place called the Bat Cave. Instructors generally left the candidates alone in the Bat Cave, but only 30 minutes was the allotted time for meals and snooze breaks.

Deklan's First Phase of SEAL training unfolded with unspeakable and constant pain. Failure was the reality to face everyone who predicted failure. Could a lifetime of regret be less punishment than five more weeks of what he was experiencing? The daily grinder PT pain sessions were only the tip of the training iceberg. Phase One was a 24-hour character examination where the slightest regression was designed to weed out the weak candidates. SEAL training character drills included surf conditioning, drown proof testing, the grinder, surf passage (boats overhead), log PT, the obstacle course and running up to 20 miles per day with loaded ruck sacks. All the tortuous physical training sessions led into the final stage of Phase One and Hell Week.

> *** Breaking BUD/S...how regular guys become Navy SEALs written by D.H. Xavier:
>
> When a man is contemplating suicide, he may have a lot of cares and distractions. When he decides, finally, that he's going to do it and that he is dead already, he experiences a lightness of being – a freedom from the most basic cares. Psychologically, beginning Hell Week is similar to committing suicide because the choosing has been done and all that remains is the act of Hell Week. And all you have to do is buckle down, do it and not quit. And what other chance will you ever have to compete balls-to-the-wall for a week straight? BUD/S training and Hell Week in particular simply force you to pay in pain up front for a lifetime of respect. I'm excited thinking about it. You should be too.

. . .

For Deklan and virtually every SEAL candidate that survived Hell Week, the five plus days were the most difficult days in their lives. Hell Week began on a Sunday. Deklan and his classmates were brought into a classroom near the quarterdeck. The entire remaining class was ushered into a large classroom filled with pizzas and water bottles. The class knew that Hell Week would begin that day, but no one knew what time the torture would commence. By the end of Hell Week, almost 80% of all SEAL candidates were gone. When the class first arrived, the group was proud they had made it to Hell Week and enjoyed the pizza that had been laid out. The apprehension hung in the room like cigar smoke. Nervous idle chatter and scanning eyes filled the room as the time dragged on. Candidates dozed off or simply imagined that maybe they would be spared the horror of Hell Week. Some candidates daydreamed from their ongoing state of physical exhaustion that perhaps the massacre would not come at all. Time continued to drag on.

Finally, when half the candidates had dozed off, the double doors burst open. The nighttime filled the room immediately. SEAL instructors screamed at their candidates to get the fuck up and outside. Blank rounds were fired from automatic weapons. More than a dozen SEAL instructors lined the exit route from the classroom to the beach. Machine gun fire filled the night sky like an enemy surprise attack had been launched on the Silver Strand. San Diego residents were always made aware of the start to Hell Week. M240 machine gun fire lit up the chaos like a series of blinding strobe lights. Flash grenades were thrown into the room. M-80 small arms explosive sticks were lit and thrown into garbage cans inside the room to simulate further an enemy attack.

The chaos was nothing short of a war zone. The group was ordered to the surf and then to the sand. The orders came from rows of instructors with bullhorns. Orders were blasted through the bullhorns to run double-time down to the ocean and dive into the midnight water. Candidates were ordered to the sand after the icy

CHAPTER NINE

cold ocean swim. Candidates were forced to roll in the sand until every inch of the body was covered in sand and then the candidates were ordered to run back to the grinder for relentless rounds of grueling calisthenics. Deklan was forced to perform thousands of push-ups, pullups, and butterfly kicks. The ocean procession called wet and sandy was repeated for hours. The chaos and the small arms fire seemed endless. The bell near the main office kept ringing. Deklan was near a physical shutdown and delusional from exhaustion but was certain he had seen a line of SEAL candidates waiting to ring the bell on the first night of Hell Week. More than 120 men began Hell Week. 42 men finished Hell Week. One class, years before, had a graduation rate of zero. Every candidate rang the bell.

Deklan took his time math calculations to Hell Week and that may have been what got him through the week. Deklan figured out there were nearly 128 hours of Hell Week. No one was allowed to wear a watch, but Deklan found every chance to look at a clock or a watch on the wrist of one of the instructors. When Deklan was getting yelled at on the grinder during excruciating PT sessions during Hell Week, he always managed to look at the watch on his tormentor. Everything had a 51% level to reach. Once a task was 51% complete, Deklan figured he was home. 51% of the week. 51% of any task until the next meal. Deklan convinced himself that once a task was more than 50% complete, then quitting made no sense. Sleep deprivation often caused reasoning to vanish, and Deklan fought hard to hold onto a few thin shreds of logic that served him well. The staff was required to feed the candidates eventually, they were required to allow some of the candidates to become Navy SEALs because there were Navy SEAL Teams, and the instructors didn't want anyone to die…became Deklan's missionary sacred prayer. Time was incremental at best and had to be viewed in small windows. Hell Week was a mindset not a dick-swinging contest to find out who was the toughest candidate. Teamwork determined success during Hell Week, not strength or athletic prowess. Death had to be the only reason to stop. Once death had been selected as one of two options, then clarity emerged. For Deklan, he was going to succeed or die before he quit the program.

Once that was the plan, Deklan realized the chances that he would die were not high. Therefore, viewing the circumstances through the few thin shreds of logic that remained, Deklan surmised that he had a better chance to get through Hell Week than he did of dying during Hell Week.

Deklan came the closest to quitting during Hell Week on Wednesday night. That was the first sleep offered to the class after three days awake. Deklan was afraid to go to sleep because they were going to wake him up in seconds. After urinating in his pants because he was too tired to get up, Deklan finally fell asleep, only to be awakened in what seemed like two minutes. The trek back over the sand berm to the ice-cold ocean was an 800-yard torture march. Men believed they were running but most of the men were barely able to walk in tiny steps. Open skin wounds covered Deklan's entire body. Deklan's legs and groin area were raw, and each step was excruciating. I can fucking quit now, Deklan thought, but didn't. Once back in the salt water, the pain doubled but he had been there before and therefore, he could do it again. Deklan entered boot camp weighing 205 pounds. Deklan entered BUD/S weighing just under 200 pounds and finished Hell Week at 170 pounds. Thursday night of Hell Week, the remaining recruits were asked to paddle their boats all night long in "Around the World" races along the Strand in the open ocean to the elephant cages and back to the base. On Friday, the last day of Hell Week, trainees were instructed to "Hide the Grapes" submerging their bodies and shaved heads over and over in the mucky waters until they were told to come up.

Deklan Novak had no idea what kept him from ringing the bell. The mind found pain management levels that appeared unattainable and impossible to achieve. Hell Week forced the class to endure hours in the freezing waters off the Coronado coast amid the sewage from Tijuana, Mexico. Guys completed Hell Week with their bodies in massive stages of breakdown. Candidates completed Hell Week with E. Coli, pneumonia, severe diarrhea, broken bones, stress fractures, severe dehydration, Mesenteric ischemia, muscle tears and hundreds of open exposed abrasions. SEAL candidates regularly hid medical

CHAPTER NINE

issues that could have cost them their lives or the chance for a normal life outside of the Navy. The SEAL candidates had not signed up to live a normal life. They signed up to become Navy SEALs. Once SEAL candidates had reached Hell Week, completion was closer yet no more assured than the day they had arrived in Coronado.

When the last day of Hell Week arrived, Deklan and the remaining 42 SEAL candidates from Deklan's original class of 274 men were struggling to remain upright, they were exhausted, covered in sea bacteria, sand, abrasions, and they carried infections from head to toe. Many were disorientated and struggled to take small steps. Deklan's close friend, John Cook, during BUD/S had four stress fractures in each leg following Hell Week and finished barely conscious. John Cook received an injury roll back to recover. John was held back from Class 264 until Class 266. Cook was allowed to join Class 266 after Hell Week, since John had already survived Hell Week.

On the Friday morning of Hell Week, the remaining SEAL candidates knew the Class would be secured soon, but they didn't know when. There was an eight-mile hike back to the quarterdeck from the beaches where the overnight paddling excursion ended. Tables had been set up with food and beverages and the dress white officer's uniforms milling about signified something was about to end.

"Class 264 secured." The Admiral's announcement came down from the grinder. Deklan had survived Hell Week. Pizzas were laid out for consumption and the weekend was reserved for medical checks and recovery. Walk week followed Hell Week. The SEAL candidates were still required to muster and were required to spend more time in the water for PT during walk week than hammering out thousands of pushups, pull-ups, and flutter kicks on the quarterdeck. The teams were allowed to walk to their meals for the one week following Hell Week and everyone received extensive injury care and medical attention to address the 5.5-day torture marathon. Phase One was nearing completion.

Deklan lined up for the 5.5 nautical mile swim in the bone chilling waters along the Pacific coastline of Coronado. The grueling ocean test took place not long after Hell Week. Battered from 12 weeks of

BUD/S preceding the swim, Deklan and his classmates prepared for the four hours plus swim test. The bacteria in the water stemming from raw sewage that traveled from Tijuana was more dangerous than the Great White sharks patrolling the waters like invisible Grim Reapers. If the chafing, sharks, seasickness, currents, and cramping did not sink the candidate, hunger often caused a loss of consciousness. Burning thousands of calories in the ocean was never a good precursor to swallowing a massive amount of sea water. Deklan and his classmates stood on the beach for inspection before the swim. Navy SEAL Instructor Kurt Dolan stopped in front of Deklan Novak.

"What is this, Iceman?" Senior Chief Petty Officer Dolan shouted at Deklan Novak, using the nickname that the Phase One instructors gave Novak. Iceman was Chuck Liddell's nickname in the UFC. Photos of Novak's brief fighting career and his unique Mohawk-hairdo victories had appeared on the base. SCPO Dolan was no more than four inches from Deklan's face. Kurt Dolan was a mountain of man, afflicted with an inbred affection to inflict incessive pain on the SEAL candidates, well above his appointed mission statement. Dolan was a block-headed Irishman, with veins always popping from his neck in anger. Dolan wore full sleeve tattoos on both arms and carried a violent, hair-triggered temper that was more prevalent in the Pelican Bay State Prison gang mentality than it was inside a strict military background based on following a chain of command. SCPO Dolan reached down and lightly touched some Vaseline that had oozed down from Deklan's arm under his wet suit.

"Sir. That is Vaseline. Sir." Deklan replied in defeat, knowing he was busted. Many shortcuts and tiny passages of wisdom had been passed down from class to class. The shortcuts only worked if you got away with them. Vaseline under a wet suit was a banned maneuver that helped against the ocean temperature.

"Since the rules here do not apply to you, asshole, take off the wet suit, Iceman. Then go get wet and sandy, Novak. Then, put the wet suit back on." SCPO Dolan barked.

"Sir, yes sir." Novak barked out and did as he was instructed. His body would be covered in sand before his wet suit was put back on.

CHAPTER NINE

"Sir, yes, sir what asshole?" Dolan persisted. "You agree that the rules do not apply to you?" Dolan was screaming and the 41 remaining SEAL candidates watched.

"Sir, no, sir." Novak yelled back. "The rules do apply to me, sir. Hooyah, sir."

"Shut the fuck up, asshole." Dolan shot back. "I didn't tell you to elaborate on anything did I Iceman?"

"Sir, no, sir." Novak gagged out. The day was not progressing well. Deklan Novak ran down to the surf and got wet and sandy for the 5.5-mile ocean swim. Novak was sent back three times to the surf to gather more sand on his body. SCPO Dolan noticed a tiny flesh-colored spot behind Novak's right ear that was not covered in sand. Back to the surf. Nolan took a small shovel filled with sand and shoved the sand down Novak's back into his wet suit by hand. Dolan forced more sand down the front of Novak's wet suit. Novak was told to remove his hat two more times to allow more sand to enter his eyes, nostrils, and mouth. Another visit back to the surf for good measure and then the swim began.

The swim was a tortuous exercise in tearing Deklan's flesh apart with each stroke in the water. Extreme chafing was a normal occurrence during the 5.5-mile swim test for most participants. When the body was covered in sand underneath a wet suit, the unrelenting friction against the skin left Deklan's entire body a floating mobile triage unit. Added to that was a constant torrent of salt water and Deklan honestly believed he was going to die. At the end of the swim, Deklan had a core body temperature of 92 degrees, couldn't stand up from the raw flesh and the weak state of his body. After 24 hours under medical care and a hot tub to raise his core temperature, Deklan was close to being declared medically unfit to continue. The halt to his training would have sent Deklan back to another class. Novak would have had to recover physically and wait for the next class to get through Hell Week. Since Deklan had survived Hell Week, his re-entry into another class would begin following their Hell Week. Being sent back on an injury roll back was not the same as being dropped out of the system for performance roll backs, but it was a major setback for how far Deklan had come. For a SEAL candidate to get through Hell Week only to get rolled back was gut-wrenching.

SCPO Tyler Dolan killed the medically unfit status and sent Deklan back to his class. Deklan could barely walk back to his barracks, each step an excruciating reminder of the wet and sandy five-hour torture session. SCPO Dolan was standing at the entrance to the barracks when Deklan arrived from the medical care unit.

"Hooyah, sir." Deklan's graveled voice was barely heard as he passed SCPO Dolan.

"Fuckin A, Iceman."

Deklan Novak Survived Phase One. Phase Two was the combat diving phase and Phase Three was land warfare where SEALs were introduced to their weaponry. SEAL candidates did not hold or fire weapons in Phase One and Phase Two of BUD/S training. Airborne School and SEAL Qualification Training (SQT) followed Phases 1, 2, and 3. SQT was a 26-week training regimen designed to take students from the basic skills mastered and learned in First, Second and Third Phases to the advanced levels of Naval Special Warfare necessary for the SEAL level of tactical training. After SQT, the SEALs received their Tridents.

Deklan Novak had an issue in Phase Two. Deklan had organized his training time on the weekends to work on what he considered were his weakest events upcoming. Deklan and a few guys would take hours on each weekend to work on things like the Obstacle Course techniques or while in the combat diving second phase, the men worked on long distance underwater techniques and various swim drills. Mastering the techniques made the exercises easier and shaved time from the clock. The rope swing was near the end of the O-course and the body was beat-up. Knowing how to maneuver the ropes, for example, was as important as the strength to handle the ropes. In Dive Phase, Deklan did not spend much time on the Tread Test. Candidates were dropped into the pool with twin scuba tanks (80 lbs.) and a 12 lb. weight belt attached to their waist. The men had to keep their hands and head above water for five minutes without dropping their hands

CHAPTER NINE

or allowing their heads to go below the surface. Deklan figured that test was not going to be the hardest test on the planet.

When the day of the Tread Test came, the SEAL candidates were instructed in the parking lot on how to put on their tanks. The task of putting on the tanks was timed. The task was required to be accomplished blindfolded as well, to simulate putting on the scuba tanks in black water unable to see and the men had to be able to help their teammates put on their tanks in darkness. Everything was timed and recorded before the pool test began. The impending difficulty of the Tread Test had begun to creep into the mind of Deklan Novak. The images in his head briefly took Deklan back to his childhood when he got caught stealing two Street Fighter video games from a Target store in Deerfield, Illinois. When a security guard grabbed Deklan outside of the main entrance and confronted Deklan with an inquiry on what was under his jacket, Deklan realized that he had not prepared properly for this unlikely, yet entirely possible scenario.

The Tread Test began. The men were not allowed to wear fins. There was a group of SEAL candidates relaxing next to the pool. They had already passed the Tread Test and were invited by the Phase Two instructors to watch the others. Smiles, laughter, and taunting accompanied Deklan to the edge of the pool. The happy gallery had practiced the Tread Test repeatedly. They passed on their first attempt. Deklan had not practiced the Tread Test at all. How hard could it be to stay afloat for five minutes? Each candidate was inserted into the pool with one instructor next to him. Deklan began his test. Within 1.5 minutes, Deklan was slipping under the water. Deklan could not breathe. The weight of the tanks and the belt sent him underwater, and he fought like hell to grab any air whatsoever. Immediately, it became evident to Deklan that he was not going to pass the test. With the realization of failure, Deklan swam to the side of the pool to extricate himself from drowning. As he reached for the edge of the pool to pull himself out of the water, Deklan's instructor pushed him back to the middle of the pool, away from the edge and the ability to intake air.

"Oh no, my friend." The Phase Two instructor smiled. "You are out here for five minutes regardless. I don't care if you drown."

"Sir, yes, sir." Deklan barely got a sound out from his water filled lungs.

"Get your hands out of the water, now!" The Phase two instructor barked.

"Sir…" the words never made it out.

"I said, get your hands out of the water, now!" The instructor's decibel level breached the water and Deklan heard everything. Deklan slipped back further under water and felt like he was about to die. Deklan tried to get to the side of the pool again and was again pushed out to the middle of the pool by the Phase Two instructor. Deklan was certain that he was going to die. Finally, Deklan decided to fake his death and that would force the staff to get him out of the water. It was the last chance he had to survive the ongoing nightmare. Deklan assumed his SEAL days were now over, but he did not want to die in the pool.

Deklan went still and the weights with the tanks pulled him straight down. Deklan left his eyes open but did not move them. He burrowed straight to the bottom. The instructor followed closely. The instructors wore weight belts as well to navigate any emergencies like the one Deklan was faking. The Phase Two instructor swam next to Deklan and got up close to Deklan's face. With his eyes open, Deklan could see the instructor laughing and shaking his head. The charade failed. The Phase two instructors were so adept at what they did, they allowed many candidates to pass out and hit the bottom of the pool. They could raise the man up in seconds and revive an unconscious candidate immediately. Navy SEAL instructors walked a very fine line between lessons and caution. The lessons won hands down. Deklan's impersonation of a dead guy gave the crew a huge laugh and when Deklan finally was pulled from the pool, choking, crying, and spitting phlegm wads of snot from his facial orifices, the announcement came loud and clear over the loudspeaker…Novak failed. Deklan Novak was beaten. At that moment, he vowed to conquer the Tread Test if given another chance. Deklan knew he would practice the test until he could do it in his sleep. Unfortunately, the Phase Two instructors made Deklan take the test two more times that same day and Deklan failed miserably each time.

Deklan Novak had to go before a Navy SEAL board to determine his next assignment. The performance rollback was the only blemish on a good candidate. The board reviewed the entire BUD/S training record for Deklan Novak. Novak had completed Hell Week. Novak had excelled on the Strand and the board decided that Deklan would be able to return with Class 266. Deklan's original class was 264. Deklan needed time to practice for the Tread Test and heal a shoulder injury that had lingered since the MMA through BUD/S and into Phase Two.

Deklan crushed the Tread Test with Class 266. Novak completed Phase One, Two, and Three, Airborne School and SEAL Qualification Training (SQT). Deklan Novak received his Trident. There was a graduation ceremony that was regal. The party that followed with Novak's unconventional family in town became a donnybrook that resulted in several arrests, an expulsion from a San Diego hotel and more than one trip to the emergency room at Sharp Coronado Hospital.

The last thing Deklan heard from the graduation white tent on the afternoon of his Trident ceremony was a message from the Commander at Naval Amphibious Base Coronado in a brief speech to the newest Navy SEALs. The message closed with the following line:

When men of our ilk are no longer needed, there will be no more war. But that time is not now.

CHAPTER TEN

2007...THE WEEK BEFORE DEKLAN NOVAK RECEIVED "THE BUDWEISER" ALSO KNOWN AS THE SPECIAL WARFARE INSIGNIA "SEAL TRIDENT"

THE SEPTEMBER 11 attacks of 2001 caused the deaths of 2,996 victims and 19 hijackers who committed murder-suicide. Thousands more were injured. Most of those who died were civilians. There were 343 members of the New York City Fire Department and 71 law enforcement officers from the New York City Police Department who died on 9/11. There was one United States Fish and Wildlife Service Office of Law Enforcement officer who died when United Flight 93 crashed into a field near Shanksville, Pennsylvania, 55 military personnel who died at the Pentagon in Arlington County, Virginia and the 19 terrorists. More than 102 countries lost civilians citizens on 9/11/01. Charlie "Cobra" Coletti re-enlisted in 2002 after having served nine years (1989-1997) as a Navy SEAL.

Less than a week remained before Deklan Novak was going to receive his SEAL Trident and Deklan crossed the Coronado Bridge that spanned the San Diego Bay wondering how he had made it to the finish line. How much of getting through BUD/S, Airborne school, and SQT was seeded in the doubt that everyone threw his way? What did Deklan Novak have to prove to anyone? Had Novak proven anything yet? One man had a profound effect on Deklan and Deklan wondered if he would ever see SCPO (E-8) Charlie "Cobra" Coletti again. SCPO Coletti retired after 18 years as a United States Navy SEAL a week before Deklan Novak received his Trident.

SCPO Coletti got his nickname from the 1986 film starring Sylvester Stallone as Lt. Marian "Cobra" Cobretti, a member of the elite "Zombie Squad" of the Los Angeles Police Department. Like Stallone in the film, Coletti wore his aviator sunglasses as if they were attached surgically to his face. There was a toothpick sticking out of the right side of Coletti's mouth during every waking hour. Coletti carried an endless supply of toothpicks because so many of them came flying from his mouth during the animated instructions extended to the SEAL candidates during BUD/S. SCPO Coletti was a BUD/S instructor on the final leg of his SEAL career, not a popular position for most retiring SEALs but SCPO Coletti was not like most men. Coletti stood 6'1" and weighed 250 hard pounds. Coletti was 40 years old. Coletti was married once and divorced during his second stint with the SEAL teams. SCPO Charlie Coletti rode Novak hard from Day 1 at BUD/S and especially during Hell Week, yet Coletti was the reason Deklan never quit. SEALs ran an estimated 150-200 miles during Hell Week. Many of those runs included a 70-pound rucksack attached to the candidate's back. Since Hell Week was 5.5 days in duration with virtually no sleep, SEAL candidates during Hell Week ran between 27 and 40 miles per day. Competitive marathon runners over many decades averaged 10-12 training miles per day.

Charlie Coletti became a Navy Seal in 1989 and served until 1997. In 1997, Coletti left the Navy and enrolled at Ole Miss, not far from his Jackson, Mississippi hometown. The 2.5-hour ride between Jackson and Oxford, Mississippi was perfect for a quick visit home to see his mother and close enough for her to come and watch her son play football for the Rebels. Charlie Coletti, at 27 years of age, walked onto the football team as a defensive tackle under Tommy Tuberville and played for two years. Coletti also won the 1998 NCBA (National Collegiate Boxing Association) heavyweight boxing title. "Cobra"

CHAPTER TEN

redefined toughness among the men that set the bar on the discussion.

After the 9/11 attacks, Charlie Coletti rejoined the SEAL teams, serving on SEAL Team One, Four and Six. The Naval Special Warfare Group, also known as DEVGRU or SEAL Team Six was the most elite counter-terrorism unit in the Navy. Coletti's assignment to the Silver Squadron inside SEAL Team Six signified his role on direct-action missions or close-quarters battles (CQB). SEALS had to apply to become a member of SEAL Team Six. Other SEAL Team assignments were directed after SQT. Applications to SEAL Team Six were unilaterally rejected if the SEAL applicant was married. Charlie "Cobra" Coletti was not married during his time with ST6. As the end to his SEAL career approached, Coletti disdained anything administrative and asked to be placed on the BUD/S staff for his final assignment in the SEALs. Having a hand in the training of men going to war was a greater responsibility than going to war.

"Immortality is what we can pass on to the next generation." Coletti often told his SEAL candidates at BUD/S. "Our job here is to show the world what we got. You either step up here and earn the right to show the world who we are or step aside quickly."

Hell Week for Coletti was an artist painting his finest work. Charlie "Cobra" Coletti didn't assign punishment during Hell Week for candidates that couldn't keep up, "Cobra" participated in the punishment he doled out. Candidates had to be taught to think about their teammates first and before anything else. When one sleep-deprived, exhausted SEAL candidate failed to think of the team first, miles were added to rucksack runs in the Coronado sand for the entire six- or eight-man Hell Week team that was home for the one selfish candidate. Coletti participated with the punished candidates. The punished team not only had to run additional weighted miles, but they also had to watch their 40-year-old instructor struggle as they did. Guilt with the candidates not only cemented home the concept of teamwork but illuminated the necessity for sacrifice in battle from the highest ranked teammate down to the E-1. The pain of Hell Week had an end, ring the bell, or learn how to put the team first. At war, the end was going home safe as a platoon or kneeling over the remains of a fallen teammate before the C-130 Angel flight to Dover AFB.

During Wednesday night of Hell Week, Deklan realized he could barely walk. The chafing on Deklan's inner thighs had turned from painful yet superficial abrasions into full-blown open skin wounds and raw flesh. Bacteria from the contaminated waters of San Diego Bay had infected many of the exposed wounds. Deklan's groin area, which resembled a science project on flesh-eating parasites, was similarly exposed and the constant running in fatigue pants and service-issued combat boots made each excruciating mile a progressive torture trek.

"Quit, Novak. I know you are not going to make it through this week." Coletti spoke to Deklan immediately after the abbreviated break for a meal had ended. "Look at you man. You cannot walk and we have a ton of running to do today. Today is only Wednesday, boy. Do you know how many hours of this are left?"

Sir, no, sir." Deklan Novak managed to retort through the remnants of his food falling from his mouth onto the filthy shirt stuck to his frame. The excruciating pain associated with remaining upright caused Deklan's legs to buckle, twice.

"There are more than two days left, Novak. You cannot stand. You cannot walk and you sure as fuck cannot run. You're done Novak." Coletti announced and started to walk away.

"Sir, no, sir!" Novak struggled to shout but the message was vocal enough. "This is all I'm going to be and all I ever wanted to be. Don't take this from me now, sir." Deklan Novak had managed to stand and was screaming at his instructor, which in real-time amounted to a barely audible begging session. The tears were undetectable amid the half-eaten food, sweat and sand stuck to Novak's face.

"Novak, you determine the outcome here, not me." Coletti was speaking to Deklan Novak with no one else in earshot. Coletti liked Novak, but that didn't enter the mindset of the veteran Navy SEAL instructor. "Do you want to know what the worst day of my life was Novak?"

"Sir, yes, sir." Deklan replied and hoped like hell that any exchange would buy him a few more minutes rest.

"I was in a helicopter crash in Afghanistan on our way to a mission that had been green lit specifically for our team." Coletti recounted. "I was on my third deployment, and we were getting high value assignments almost every other day. The Boeing CH-47 Chinook my squad took that day was hit by an RPG out of nowhere as we were trying to reinforce a Joint Special Operations Command unit of the 75th Ranger Regiment in the Helmand Province of Afghanistan. We lost two teammates that day. Four other members of my squad and two Afghan Army partners were injured and could not walk on their own. I had no clue if we were going to be ambushed from both sides or where the RPG had been fired from. I heard the M240 Bravo machine guns shooting at our position and knew we were under attack from the Taliban using our abandoned or stolen machine guns. There were no reports ahead of time on enemy positions that we would encounter prior to the target area. Do you follow me, Novak or are you going to fall asleep on the ground, boy?" Coletti was not screaming, Coletti was screaming. Wasn't he always screaming? Novak had entered a space filled with sleep-depriving delusions and hallucinations. Deklan's legs buckled again slightly but he stayed upright and kept his head up.

"Sir, no, sir" Deklan answered.

"Sir, no, sir what Novak?"

"Sir, no, sir, I am not going to fall asleep."

"You better not fall asleep Novak. Ring the bell or find your balls right now, SEAL Candidate Novak." Coletti ordered.

"Hooyah." Novak coughed out a response.

"Do you know what I did after we got shot down, Novak?" Coletti asked again but did not wait for another response. "I called back to base camp for orders on how to proceed. Do you want to know what base camp told me to do, Novak?"

"Hooyah, sir."

"The answer was "Charlie Mike." Colatti waited.

"Sir, yes, sir...Charlie Mike."

"Were they referring to me Novak? My name is Charlie."

"Sir, yes, sir." Novak responded and hoped the correct answer was what he said although the entire exchange was a blur.

"No, they were not referring to me." Coletti recalled. "For us, Novak…Charlie Mike meant Continue Mission. So, we did."

"Hooyah, sir." Novak looked up at his Hell Week instructor.

"Continue Mission, Novak." Coletti stood up erect and walked away from Deklan Novak. Deklan Novak watched a man who had served on SEAL Team Six move away slowly. There would be the first chance to sleep upcoming later that day on Wednesday. The nap lasted what seemed like seconds. The open wounds were still there. The movement continued to send torrents of pain through Deklan's legs and groin. The trek back to the ocean suddenly didn't appear impossible. All Deklan had to do was to continue. He did. Charlie "Cobra" Coletti had landed in Mississippi a few days before Novak received his Trident. Cobra was at McKenna Ranch in Pachuta, Mississippi hunting whitetail deer when Novak wore his Trident for the first time.

CHAPTER ELEVEN

2014...CARMEN GALLARDO

THE SINGLE-WIDE TRAILER was located towards the back section of the Dream Island Mobile Home Park. The two-bedroom home where Raya Nolan Thomas lived with her boyfriend Carmen Gallardo and her six-year-old daughter Kaley was 984 square feet of living space divided into two bedrooms, a living area/kitchen, two tiny bathrooms and a small closet next to the kitchen area. The mobile home mortgage had been paid in full by Raya's mother before she had passed. Mobile home parks are landlords who rent the land where the trailer homes rest. Raya's monthly rent was $975, roughly double the national average for mobile home park rental fees because Dream Island Mobile Home Park was located within the heart of the national ski resort community, Steamboat Springs, Colorado. Raya was at work, waiting tables at the 8th Street Steakhouse, a ranch to table experience in downtown Steamboat Springs where the customers selected the hand-cut steaks or seafood and then were assisted by staff in cooking their own meals on one of two 20-foot lava rock grills. Carmen Gallardo was seated at the bar a block away from the 8th Street Steakhouse inside Sunpie's Bistro on Yampa Street. Kaley was home alone, locked in the small closet next to the living room/kitchen inside the trailer that sat on the back quadrant of Dream Island Plaza.

Carmen Gallardo served two sentences at Colorado State Penitentiary in Canon City, Colorado. Canon City is a few miles southwest of Colorado Springs. Gallardo was convicted of selling cocaine with intent to distribute on his first offense and sentenced to four years incarcerated and four years of supervised release. Gallardo's second stint at Colorado State Penitentiary was for third-degree assault. Gallardo received a one-year sentence because he was a convicted felon and nearly beat a man to death. The dispute took place outside of a bar in Avon. Gallardo claimed he was only defending himself. Witnesses painted a different picture that involved an exchange of money and drugs that had gone wrong. There was an ounce of cocaine found on the victim, unconscious with a broken jaw and four broken ribs. Gallardo was arrested at the scene and had $2,000 cash on him. Third degree assault in Colorado carried a maximum of one year in jail. Gallardo pled not guilty, was convicted, and did the time.

Carmen Gallardo was a good-looking man, a vagabond hustler who avoided work like a cowboy avoided fences. If Gallardo fell into the river, he'd naturally float upstream. Gallardo stood 6'1" tall and weighed 198 pounds. Gallardo was born in San Juan and his heritage was Puerto Rican and Spanish. Gallardo was 34 years old, born in 1980. The jet-black hair was thick and always longer than the prevailing short trends that most men adopted once they joined the great American work force. Gallardo never joined the great American work force, and no one ever accused Carmen Gallardo of working harder than a dog, herding cattle in a thunderstorm. Gallardo was clean shaven most of the time and his eyes bore the uneven allure that spoke of danger and mystery. Gallardo resembled Benicio Del Toro dressed like he was rejected for a part in Cocaine Cowboys. Gallardo wore a Paul Hogan hat from Crocodile Dundee, a linen white shirt, and Sonny Crocket grey linen flat-front pants. Carmen Gallardo looked more ready to pilot a cigarette boat in South Florida than he appeared to belong to a ski town in Colorado. Gallardo would have been less conspicuous if he carried a sign advertising his products.

CHAPTER ELEVEN

Gallardo was still selling grams and ounces of cocaine, but his days were limited as the Steamboat small-time mule. Raya Nolan Thomas and her daughter comprised the ticket Gallardo had been waiting for.

Sunpies Bistro was the hangout downtown for locals in Steamboat Springs. Sitting between Yampa Street and the Yampa River Core Trail, Sunpies Bistro was a hole-in-the-wall where the South and the Rockies joined forces serving a Louisiana-style menu and the strongest drinks in Steamboat. Gallardo was a welcome fixture at the bar inside Sunpies. He was a local and the staff were all locals. Part of each evening's entertainment was people-watching as the tourists wandered into Sunpies. The hotel concierges told their guests that Sunpies Bistro was the local bar where one would taste the true flavor of Steamboat Springs. The locals all made fun of the tourists decked out in expensive Canada Goose coats, fanny packs, brand new Stetson cowboy hats that didn't fit, and boots from Kemo Sabe in Vail and Aspen that were marked up 300%. Gallardo also had cocaine, which made him the Mayor of Sunpies. The long thin bar area had multiple flat-screen televisions and was covered in European football posters, photos, and team flags. The outdoor patio was an unkempt collection of eclectic tables and chairs that had seen better days. Carmen Gallardo took the call from inside the bar and walked outside to hear better. The air was crisp on an autumn night.

"Hello." Gallardo recognized the number, pulled out a chair on the patio and watched the evening hikers file by under the lights on the river trail.

"Hello, Mr. Thomas." Donald Sanders greeted Behr Thomas or the person he believed to be Behr Thomas. "I received the E-signatures via the DocuSign. Everything is set, and the policy is effective retroactive to September 1. Sorry about the delay. Is there anything else you need currently, Mr. Thomas?"

"No." Gallardo replied politely. "Thank you for your help. I'll make sure my ex-wife is aware of the update. Good-bye, Mr. Sanders." Gallardo hung up.

Donald Sanders was an agent for the Municipal Life Insurance Company based in Deerfield, Illinois. Mr. Sanders worked as the liaison between Municipal Life Insurance Company and the Northwestern University Office of Human Resources. Municipal Life Insurance Company was the underwriting firm for the life insurance plans offered to the Northwestern University employees. The policy that Donald Sanders was referring to, was a life insurance policy on Kaley Thomas with a death benefit of $200,000, significantly more than the normal child life insurance cap of $50,000. The beneficiary listed on the child policy was Raya Thomas. Behr Thomas and his ex-wife Raya were current clients of Municipal Life Insurance Company. Behr Thomas had a $500,000 life insurance policy with MLIC with Raya and Kaley as the beneficiaries listed on his policy. The child policy taken out on Kaley was possible because the parents were existing clients and had a life insurance policy for at least double that for the child. In Illinois, the courts required the spouse assigned to make alimony and child support payments to carry a life insurance policy with the ex-wife/mother as the beneficiary. In Behr and Raya's case, the life insurance policy already existed. The court required the policy to remain intact. Carmen Gallardo impersonated Behr Thomas to take out the child life insurance policy on Kaley. The transactions were carried out via email and DocuSign.

Since Kaley had been ill, Raya Thomas applied for single parent Social Security Disability Income (SSDI) based on Kaley's constant visits to the hospital without a successful diagnosis. The SSDI paid the parent, and the child based on the application process. Kaley was awarded a monthly income of $1250, and Raya was awarded a monthly income of $3781. Raya's income from the 8th Street Steakhouse was paid in cash and kept under the table. The restaurant avoided taxes on the income and Raya avoided showing the income to Social Security. Gallardo supplied the steakhouse manager with blow and the cash payments to Raya suddenly became possible. Gallardo convinced Raya to partner up on a scheme to defraud the SSA for SSDI payments. Gallardo hadn't told Raya about the new child life insurance policy that was taken out based fraudulently on Behr's forged signature and Raya's coerced signature. Gallardo paid for the first year of the child life insurance policy upfront. Raya signed the

DocuSign email pages because Carmen Gallardo had told her the documents were a part of the ongoing SSA agreements and Raya was so coked out at the time, she would have signed anything to smoke another pipe of crack cocaine.

There was a small rectangular space cut out of the closet door up near the top of the door. The space allowed air to get into the closet and enough light so Kaley could see clearly. The door was always locked when Kaley was alone in the trailer. Gallardo and Raya did not take Kaley to daycare and never hired a babysitter to watch Kaley when they left. The couple locked Kaley inside a closet when they left the trailer. The closet had a bucket for Kaley to use as a toilet and the closet contained a folded up filthy blanket that Kaley used as a bed.

Gallardo's plan was coming together. Carmen Gallardo was not a human being. Gallardo represented the banality of evil, a man who reveled in his ability to torture under the pretext of survival. The first part of the plan began after the pair hooked up following Raya's divorce and her move back to Steamboat Springs. Carmen Gallardo and Raya Nolan had been an item during Senior year at Steamboat Springs High School. Raya moved away after high school and attended college. Carmen began a career that allowed him to join the National Crime Information Center (NCIC) as a bona fide fingerprinted convicted member of the national database for criminals. Raya met a football player and got married. Carmen was allowed to tour the Colorado State Penitentiary for an extended stay. They ended up back in Steamboat at the same time. Carmen heard Raya had moved back to her mother's place. Gallardo came by the trailer park to pay his respects after hearing about Raya's mom passing. They began seeing each other not long after reconnecting. Carmen Gallardo began the first part of his plan as soon as they started dating again.

Crack is a form of cocaine that people smoke. The term "crack" refers to the crackling sound the crystal makes when it is heated to smoke. Crack is a freebase form of cocaine hydrochloride that is processed using water and either ammonia or baking soda, until it forms a rock crystal that can be smoked. People typically smoke crack by heating it in a glass pipe; however, they may also add it to a cigarette or a marijuana joint. Crack intoxication follows three phases. The first phase of cocaine intoxication is the early stimulation of the brain, followed by a second phase of hyper-stimulation with possible toxic convulsions, tachyarrhythmia (rapid or uncontrolled heartbeat) and/or dyspnea (labored breathing). The third phase of cocaine intoxication is depression of the central nervous system. Addiction to freebasing cocaine can be instantaneous or gradual, based on factors such as purity levels and the user's genetic tendencies towards dependency.

Cocaine and crack are powerful stimulants that provide the user with a euphoric high and increased energy. Crack is much stronger than the powder form of cocaine and reaches the brain much faster than snorting the powder. The high is stronger but only lasts five to ten minutes as the brain is flooded with dopamine. The crash is hard and fast after a crack rush. Many individuals become addicted after one-time smoking crack; Raya was that user. Raya had no clue what crack was and never thought to ask. Gallardo provided crack cocaine for Raya that approached the 80% pure compound rarely found anywhere on the street. The more hands that touched the drug, the smaller the percentage of purity became to where most street level grams were around 20% pure. Raya had snorted cocaine a few times in her past and assumed Crack was the same. Addiction was not a choice. The strength of the drug erased good and evil from Raya's life. Gallardo hatched the plan to give Kaley daily doses of Lysol or Ipecac to induce vomiting and mimic symptoms of digestion shutdown. The plan to move in together, to home-school Kaley, to induce Kaley's illness and defraud the government was based on a crack addiction. Carmen Gallardo had a master's degree in drug related activities. The plan to lock Kaley in a closet and eventually kill her for the insurance money was predicated on the addiction to crack cocaine by Kaley's mother. The plan began flawlessly.

Raya had met Gallardo at Sunpies after her shift at the steakhouse. They had a drink and left for home. Kaley had been locked in the closet since 3:30 p.m. that afternoon. Kaley never made a sound in the closet, never called out for help, and never tried to escape. The trailer smelled like a dusty basement infested with dead animal remains. The kitchen sink was full of dirty dishes and the pots from the previous day's meal remained unwashed on the small counter. Raya just wanted to get high.

"Clean this shit up." Gallardo barked as they entered the trailer. "Raya, clean this up and I'll give you the pipe." Gallardo opened the closet door and looked down at Kaley.

"You remember what the rules are here don't you Kaley?" Gallardo asked the little girl who still sat on the dirty blanket in the closet.

"Yes." Kaley spoke softly.

Carmen grabbed Raya by the hair and pushed her into the wall by the front door. Raya hit the wall hard and fell to her knees. Gallardo picked her up and slammed his fist into Raya's solar plexus. The blow knocked the wind out of Raya, and she collapsed on the ground wheezing and searching for air. Raya curled up on the floor in a fetal position and tried to protect herself from any further blows to her midsection. The attacks were common. Gallardo was careful not to mark Raya's face.

"This is what will happen if you talk to anyone about our home here or what we do here." Gallardo addressed Kaley while helping Raya to her feet and then to a chair. "If you scream for help or yell to get someone's attention, then your mommy will get hurt bad, much worse than this, Kaley. Do you understand me, Kaley?" Gallardo spoke softer now and repeated the example at least once or twice a week. Kaley nodded.

"Okay. Now, come out of that closet and we can get some dinner." Gallardo announced as if they were all in agreement and things were good. Kaley crawled out of the closet and walked quickly to get past Carmen Gallardo. Gallardo grabbed Kaley's arm hard as she tried to walk past the man who changed everything. Kaley grimaced and pulled her small shoulders up to absorb the pain from Gallardo's grip. Gallardo wasn't looking at the little girl. He pulled Kaley close to his leg, switched his grip to her hair and stared down at her mother. The devil's highway was not a road less traveled in the Dream Island Mobile Home Park.

"I don't have to remind you what will happen to your daughter if you decide to share anything about our arrangement to the local law enforcement members?" Gallardo made his point again and released his handful of Kaley's hair.

Before Gallardo started to address the filthy kitchen, Gallardo gave Raya the crack pipe and a sizeable chunk of crystal. Raya was reminded of her position quite often. Raya knew that Gallardo could have her arrested for making her child sick. Gallardo was nothing if not thorough. Carmen Gallardo knew nothing about factitious disorder imposed on another (FDIA), also known as and originally named Munchausen syndrome by proxy (MSbP). What Gallardo remembered was The Sixth Sense, a movie starring Bruce Willis. In the 1999 film, Cole Sear (played by Haley Joel Osmond) discovered Kyra Collins, a female ghost child, vomiting. Cole found out who she was and went to the funeral reception at her home. In her room, Kyra gave Cole a videotape that he handed to her father. The tape revealed Kyra's mother poisoning her daughter's food with Lysol, alerting her father to the reality of her death, and saving her sister from the same fate.

FDIA and MSbP were conditions where a caregiver, most often the mother, created the appearance of a health problem. Permanent injury or death of the victim occurred often because of the disorder. Gallardo reminded Raya each day that law enforcement did not like child abusers. If Raya resisted Gallardo's child-rearing agenda, the crack pipe would be taken away and Raya would be charged with child abuse. Gallardo reminded Raya often that while the court system and law enforcement in general vehemently resented child abusers, a prison sentence was much worse. The federal inmates at any state women's prison treated child abusers harshly and believed that assaulting or killing child abusers was a service to society. Inside the societal hierarchy of prison inmates, cop killers garnered the most respect. Convicts who had committed crimes against children, especially physical abuse, were referred to as "dirty prisoners" and those "dirty prisoners" were hated, attacked and often murdered inside the prison walls. Raya wanted the crack above anything including the safety of her daughter. The addiction and the threat of incarceration was a security blanket for Carmen Gallardo stronger than Israel's Iron Dome. Raya lit and smoked the crystals in the pipe. The world relaxed.

"Get up and help me here." Gallardo barked after Raya finished her pipe. Raya didn't move.

CHAPTER TWELVE

2009...DEPLOYMENT TO AFGHANISTAN - 1ST DEPLOYMENT

DEKLAN HAD ONLY BEEN with a platoon for a few months. Deklan graduated BUD/S and was assigned to SEAL Team One Alpha Platoon led by Chief John Simmons and Lieutenant Kenneth Powell. Those two had handpicked the four new guys from the incoming BUD/S Class 266 to join the other six members of ST 1 Alpha. Deklan remembered seeing those two officers towards the end of his SEAL Qualification Training dressed in the latest multi-camouflage uniforms with their rank and trident emblazoned, snooping around and directing questions to the instructors and occasionally talking to one of the candidates. Little did the new guys know that there was a competition going on amongst the next eight platoons from the West Coast SEAL Teams, SEAL Team One, Three, Five, and Seven. There were two sister platoons in each team that deployed simultaneously, called a Task Unit. Throughout the next three-month training cycle, the eight platoons were judged in every criterion by the top brass in Naval Special Warfare (NSW) to determine which platoon would be most suited to deploy to Afghanistan attached directly to and be fighting alongside a SEAL Team Six Task Unit. That was an absolute dream for any SEAL but especially for Deklan and his friends coming directly out of BUD/S. There could be no greater privilege than to go from a BUD/S candidate to a deployment to a combat zone with SEAL Team Six in under a year. Deklan went all in to make that dream come true.

In the middle of the chase to fight alongside the SEAL Team Six Task Unit, is when Deklan's younger brother DJ died. Just a few months after graduating BUD/S, Deklan's momentum and mantra was full steam ahead to guarantee the desired landing spot was selected as his destination, SEAL candidate Novak received the fateful call from Dana about his brother. The call tore open new wounds that Deklan knew would never fully heal. The chase to land with his desired deployment took on an entirely new dimension. Losing his younger brother, drove Deklan into a single-minded march to excel as a warrior beyond the high standards set before him. Deklan could not handle sitting with DJ's death, sorting through all the subsequent feelings, yet it gave Deklan great fuel to accomplish and push himself and his platoon way beyond what they ever could have accomplished. The tragic loss allowed Deklan to suppress the pain from DJ's death but allowed him to use it as a seemingly endless source of drive.

In Deklan's personal time, he mourned the only way that he knew how by shutting down emotionally and finding solace at the bottom of a bottle. Deklan could manage, hardworking and heavy drinking, but he had no coping skills for managing his relationship with Amber who was also mourning. She too had no skills to handle what she was going through let alone what they were going through as a young couple. Deklan really felt all alone at the time, and he was alone by choice. Unfortunately for the marriage, Deklan was chasing a SEAL Team Six deployment to Afghanistan to save the world while mourning the death of his little brother. Amber at that time was depressed, angry, lonely, homesick and without direction or passion. Amber's anger and Deklan's solitary SEAL journey remained present in the marriage for years. Both Deklan and Amber knew their destiny as a couple was tenuous, but children arrived, and Christian obliga-

CHAPTER TWELVE

tions held strong. But DJ's death was one of many turning points for Deklan Novak. Deklan stopped caring about many things peripheral in his life. He stopped caring what other people thought. The fragility and mortality of life emerged like the sun rising on a new morning. War had a bigger purpose. Time became real and tangible, and Deklan Novak wasn't going to waste it concerned about anything besides what he thought was best for SEAL Team One and those around the Teams. Placing the Teams ahead of his family had not been a calculated plan, yet the evolved reality that transpired left Amber and the boys adrift while watching the news to understand why they did not have a husband or a father present or one they knew very well.

Amber became increasingly paranoid as her husband was spending more time at the Teams working, training, and prepping for more work and training. Deklan didn't care. After DJ's death, Deklan realized that he could not make Amber happy, and an even more important epiphany was the realization that Amber's happiness was not his responsibility. Amber pushed back against Deklan, hard on his absence. Deklan stayed strong while their relationship grew increasingly chaotic. Deklan's SEAL Team One work excelled and the growing distance between Amber and her husband accelerated.

Deklan's platoon was clearly the front runner to receive their desired assignment. There was no way Alpha Platoon was going to let up and relinquish any of the distance they had gained between their platoon and the next closest platoon. Deklan belonged to a platoon of freaks. One SEAL in Deklan's platoon had been an ex All-American middle linebacker in the SEC. Another SEAL in Alpha Platoon had also been an MMA fighter in California. Two platoon mates transferred from Marine Recon, and yet another Alpha Platoon teammate was the most pound-for-pound freak of nature athlete that Deklan had ever met, who had broken both legs during Hell Week at BUD/S. Each leg had three stress fractures, and the man was standing at attention when the class was secured at the end of Hell Week. The platoon had an elite sniper that graduated from NYU, an explosives genius, and a chief that had the uncanny instincts of General Patton with an unmatched ability of putting each member of the platoon in the right place at the right time. SEAL Team One Alpha Platoon was unstoppable. Deklan's platoon dominated everything and every event. And they knew it. At the time, Deklan lived two lives. One life lived was as a humble hardworking member of an elite group of warriors preparing for selection and deployment to an unknown war zone filled with little-boy aspirations and combat fear, while the other life lived was as a sorrowful, constantly intoxicated, unhappy husband.

The NSW selection process dramatized the verdict to keep all the other platoons engaged, but Deklan's platoon was selected unanimously as the platoon that would represent the West Coast Vanilla SEAL Teams augmenting SEAL Team Six Red Squadron to Afghanistan. Three weeks after the selection was made. SEAL Team Six decided that they had space and wanted an additional platoon after Alpha Platoon was selected, so Deklan's platoon put forth the argument that their sister platoon Bravo would be best suited to man the SEAL Team Six request. Deklan's platoon recommendation was quickly confirmed and SEAL Team One Task Unit One (Alpha and Bravo Platoons) began gearing up for deployment, three weeks away from wheels up to Virginia Beach and Damn Neck to meet Red Squadron on their home turf. SEAL Team Six Red Squadron was the unit on May 2, 2011, that killed Osama Bin Laden in Abbottabad, Pakistan.

Information access related to a Navy SEAL workup preparing for war was limited. Deklan's marriage, already on a seismic logarithm exceeding a major earthquake warning, was further damaged by the secrecy required by USSOCOM guidelines that prevented Deklan from providing the details of his deployment and training to his own family. The SEAL Team Six Red Squadron facility was everything Deklan imagined. No expense was spared. All aspects of warfare were addressed equally and comprehensibly. Every detail related to warfare training and preparation was concocted and built by the world's most experienced and well-trained professional warriors. The complex was a hands-on training world built by men that had experienced the world in a very hands-on way. The Red Squadron war room vaguely reminded Deklan of the BUD/S quarterdeck on steroids and methamphetamines. War trophies from all over the globe adorned the walls and shelves. Deklan's heart pounded as his eyes feasted. The experience was dizzying. Most of the Red Team guys were already in Afghanistan. Two Red Team members that stayed back walked into the room. Deklan felt his eyes bug out of his head and a huge smile threatened to invade his face and expose his excitement. Deklan resisted and instead studied every minute detail of these two megalithic creatures before him. Deklan wanted to grill the two Red Team guys about the real-life dangers waiting for the upcoming deployment and how best to assimilate a seamless rendezvous with the current guys in Afghanistan. Training exercises, regardless of how difficult and intense they had been, were never a substitute for combat experience. Deklan remained silent. The Red Team members sat down in two chairs like humans not prolific monuments, crossed their legs and turned on a pre-deployment training projector.

Four days later, Novak and his platoon were flying from Virginia Beach commercially to Rammstein, Germany then transferring to a military C-17 for the final leg into Kandahar airbase, Afghanistan. Upon arrival in Kandahar, the platoon met up with the main body of Red Team guys for a few days. From there the platoons broke up and linked up with smaller crews on Forward Operating Bases near the front lines of battle. Deklan and three other guys were assigned to FOB Camp Chapman, Khoust, Afghanistan near the eastern border with Pakistan. Chapman was set up at the beginning of the U.S.-led offensive against al-Qaeda and the Taliban in 2001 and began as an improvised center for operations. A military base at the beginning, Chapman was later transformed into a CIA base and one of the most secretive and highly guarded locations in Afghanistan. The base eventually evolved into a major counterterrorism hub of the CIA's paramilitary Special Activities Division, used for joint operations with CIA, military special operations forces, Afghan allies, and had a housing compound for U.S. intelligence officers.

Over the next three days Deklan and his crew spent time training on recently updated, state-of-the-art facilities in country allowing the

Team members to check and double check all the gear and weapons systems for the last time before the teams were operational. They were issued upgraded versions of gear such as night vision goggles, weapons optics, red dot lasers, light weight ballistic vests, half shell helmets, digital camouflage uniforms, and more. The SEALs saw combat the next day.

Deklan recalled, on the last night before the three fellow new guys and Deklan left for their first deployment they partied at a legendary Navy SEAL Bar, The Hot Tuna. There, Deklan reunited with several of his BUD/S friends that had been sent to the East Coast after graduation. They drank well past the last call knowing it could well be their last night in America.

In October, 2009, a couple months detached from Red Squadron and working with an Afghan Commando Unit, Deklan Novak and eight members from SEAL Team One were working out of FOB Salerno, located in the southeast province of Khost, Afghanistan. Inside FOB Salerno were sub-camps for U.S. Special Operations Command Units, although many of the details surrounding the collaborations between SEAL operators and Afghan Commando units were not made public. The main base, FOB Salerno, had grown to house more than 5,000 servicemen, servicewomen, civilians, and contractors. The base was in a hostile location and was not far from Tora Bora, the site of the U.S. military's valiant and unsuccessful cave complex battle in late 2001 against Al-Qaeda. As the search for Osama Bin Laden ramped up quickly after the attacks on 9/11, the intel at the time had Al-Qaeda headquarters in the Spin Ghar Mountains at Tora Bora near the Kyber Pass. OBL was reported to be hiding at the Tora Bora Al-Qaeda headquarters as early as two months following the attacks in New York City. Vice-President Dick Cheney told Diane Sawyer of ABC News on November 29, 2001, that, "I think OBL is in that general vicinity." After a multi-national force of Army Rangers, Delta Force and Afghan Army soldiers pounded the mountain terrain and the bunkered down Al-Qaeda fighters at Tora Bora from December 3-December 12, a local Afghan militia commander reached a truce with Al-Qaeda to allow the Al-Qaeda fighters to lay down

CHAPTER TWELVE

their weapons and surrender. Sixty Al-Qaeda fighters were captured, but the truce gave Osama Bin Laden the opening to escape into Pakistan and as discovered in 2011, eventually to a compound near Jalalabad.

FOB Salerno was also only 25 miles north of where Pat Tillman was killed in 2004 by friendly fire. The terrain was unfriendly. The missions were fraught with danger and the valuations were, at the least, questionable related to the whereabouts of any high value target (HVT). Deklan had transferred all justification for his presence in Afghanistan into saving the lives of his brothers in arms. After leaving the base just before midnight via mounted patrol aboard two armored Toyota Hilux pick-up trucks flanked by two separate Humvee gun teams on October 7, 2009, Novak's platoon was ordered to split into two groups at 2:22 a.m. One Toyota and one Humvee were unexpectedly ordered to search one of two supposed hideouts for the suspected target. The two locations were deemed far enough apart to warrant the split platoon. The intel had the target in the area for that night only. Deklan believed that splitting a platoon into two groups was the common denominator in combat, for a disaster. Novak's disdain for splitting a combat platoon had a significant recent historical connection. As a vocal fan of Pat Tillman, Deklan kept his focus on the task at hand. Regardless of the past casualties incurred from split platoons, the Tactical Operations Center ordered Deklan Novak and his platoon to locate, capture and/or kill Kasur Halal by splitting up into two units and searching the two locations where the target had been seen.

Deklan recalled the single biggest mistake ordered from the Combined Joint Special operations Task Force-Afghanistan (CJSOTF-A) on April 22, 2004, that preceded the death of Pat Tillman, was not the choice of the platoon commander on the ground. In fact, "Black Sheep" platoon leader First Lieutenant David Uthlaut objected vehemently to splitting his platoon. When Lt Uthlaut messaged his regiment's Tactical Operations Center far away in Bagram, near Kabul, he asked for a helicopter to hoist a disabled Humvee, which had stalled the mission. There was no chopper dispatched. The A Company commander ordered Uthlaut to split the platoon. One group was directed to continue with the original mission and the other group was told to bring the Humvee back for repairs. Pat Tillman's platoon "Black Sheep" (A Company, 2nd Battalion, 75th Ranger Regiment), which included his brother Kevin, was ordered to cut their effectiveness in half while stuck at the base of a deep mountain crevice in Paktia province, a precariously vulnerable region of broken roads and barren rock canyons. The breakdown of a Humvee, the ill-fated directives from the Tactical Operations Center in Bagram, located nowhere near the insertion point and the impending financial loss of a $220,000 vehicle cost Pat Tillman his life. The ensuing friendly fire and an American M-16 rifle took Pat's life. According to the Army Criminal Investigative Command dated June 2, 2005, and the Inspector General's Department of Defense Report dated March 26, 2007, the chain of command did not make critical tactical errors that eventually contributed to the confusion within the platoon that led to the friendly fire incident that took the life of CPL Tillman. The reports assigned blame to the platoon commander and the lack of responsible participants to comply with applicable standards for investigating friendly fire incidents.

FOB Salerno had been nicknamed "Rocket City" because the base was attacked consistently. In 2008, insurgents attempted to assault FOB Salerno with a double car bomb. The VBIEDs were detonated close to the perimeter of the base and killed 15 Afghan Army soldiers. The attackers attempted to breach the base near the airfield to steal Apache gunships. Several of the 30 insurgents were Saudi's trained to operate Apache Helicopters. Nearly all the insurgents carried suicide vests. Three suicide vests were detonated when the insurgents came under machine gun fire from Afghan commandos and a team of Navy SEALs. The SEALs and the International Security Forces (ISAF) ended the attack on the base. While the Taliban claimed credit for the attack on FOB Salerno, it was likely executed by the Haqqani Network, a Taliban group linked to AL Qaeda operating mainly in Khost Province.

Two months later, another group of Taliban was observed

CHAPTER TWELVE

preparing for an attack 1000 meters from the base. Before coalition forces intercepted the advance, A10 warthogs and Reaper drones pounded the Taliban staging area resulting in multiple enemy casualties and ending the threat to the base. FOB Salerno earned its nickname "Rocket City" repeatedly. FOB Salerno was home to numerous U. S. Commands over the years including Task Force Geronimo, Task Force Devil, Task Force Thunder, Task Force Spartan, Task Force Fury, Task Force Yukon, and Duke, all airborne, infantry or parachute brigades teamed with the Afghan National Army partners, International Security Forces and U. S. Special Forces.

The 2008 insurgent attack on the base and subsequent Taliban bold moves against coalition forces at FOB Salerno were attributed to a high-ranking Taliban leader named Kasur Halal, unofficially known as the leader of the Haqqani Network, a group that oversaw the financial and military assets within the Taliban. Halal, a Tier One Target, a face-card on terror-rank cards dealt during staging exercises prior to combat missions, was reported to have been seen in the villages Deklan and his gun teams had been assigned to search. Recent radio chatter had indicated that another Halal attack on FOB Salerno was imminent. By summer 2009, Al-Qaeda and Halal had formed an alliance in the region and Salerno was always vulnerable. Salerno was completely isolated from the rest of the country by the mountains and only accessible through the Khost-Gardez Pass. Halal was believed to have been behind the attack at an FOB Salerno checkpoint in late September 2009, that killed two dozen U.S. soldiers and Afghan police officers. The Lockheed Martin RQ-170 Sentinel Stealth Drone surveillance footage following the attack indicated that Halal was close by.

Deklan Novak and his platoon had two adjacent villages to search in the darkness and then were scheduled to return to the vehicles no later than 5:15 a.m. The close location of the base to the protected valley where the villagers slept required the cover of darkness to facilitate hiding their arrival. Missions were most dangerous at dusk and dawn, where the under-equipped "hajji" did not suffer from the disadvantage of nightfall. Deklan and his fellow SEALs wore advanced four-tube night vision goggles (NVGs), at $65,000 per unit, during the darkness. The gun mounts on the Humvees and the ammo cans were manned by the Army Rangers. The .50 caliber rounds could penetrate a bank vault. The six members of SEAL Team One included two SEAL snipers to pick off movement unaccounted for in the villages. Within minutes after the platoon arrived at the first target village, the Taliban lookout sentries were loading their RPG's and alerting the no longer sleeping villagers through a series of primitive yet effective signaling. Deklan, while certain the lookouts hadn't volunteered for Istishhad (death by matyr), Novak knew the lookouts were not long for this world. Deklan often wondered, however lethal his team was, where was the strategy in disguising stealth with a red flare? Humvees and Toyota armored pick-up trucks did not run silently. Darkness provided some elements of surprise, but hardly enough to justify splitting platoons. Men making those decisions were never the men that arrived on site. Deklan barely gave the disruption of their arrival another thought.

The Taliban lookouts were extinguished with the silent sniper fire that sounded like an air gun. One by one and with the precision of a diamond cutter, the perimeter interference was silenced. The village Deklan approached was hidden neatly within the small valley of a mountain clearing. The homes were constructed of sun-dried bricks and clay. Deklan pictured bearded Afghan men dressed in pajama-like trousers, long shirts and wool hats scrambling for their Soviet-era PK or PM1910 machine guns and taking positions at every window or doorway. The platoon didn't knock on the door of each residence. Deklan was a breacher and breachers provided the open doorways. Expertly measured charges or 30-pound rams blew open the protected entryways. Platoon members followed the breacher's work and each room was systematically cleared. Flash or stun grenades were tossed into rooms with perceived threats as a less-than lethal method of disorientation to the enemy. The platoon did not speak Pashto but had been rehearsed on a few key Pashto questions to assist in the door-to-door search.

"Hands up" and "move into the main room" were the first orders announced as each room was entered and women and children were found hiding. Men were most often not afforded such verbal warnings.

Failure to respond with anything short of compliance meant the discharge of an M4A1 carbine was imminent. The Afghan commandos with each group followed behind the Navy SEALs and could ask more detailed questions of the captive audience. Specifically, the SEALs wanted to know where Kasur Halal was hiding. With each search and family disruption, Halal was nowhere to be found but a convenient stash of a dozen Kalashnikov assault rifles (Russian AK-47s), PKs and PM 1910s were discovered in one house and confiscated. The SEALs assumed the rifles were sacrificed to appease soldiers searching for the suspected target. Many of the older model Russian rifles had laminated wood stocks, many of which were cobbled back together with sheet metal, duct tape, and light nails.

Deklan and his fellow Navy SEALs were clearing the last house of the village, anxious to return to the Humvees before dawn and get their asses back out of Dodge. The family disturbed in the final house gathered in the one main room of the structure. The Afghan father was a tall thin man. Weren't all Afghan men tall and thin? OBL appeared to live in the faces and inside the eyes of all rural Sunni men. The woman and mother of the three children wore a firaq partug or the traditional Pashtun clothing. The term firaq partug referred to three articles of clothing worn by Afghan women. The chador is the head scarf. The firaq is a garment like a long skirt. The partug is a type of shalwar and is the lower garment, baggy, gathered at the waist and tied around the ankles. The bearded father and three children huddled around the woman who was staring daggers through the floorboards of the primitive residence. The children looked like a movie prop, a huddled bag of used clothing waiting to be dumped into the Goodwill bin at the local convenience store. Their eyes were locked down as well, watching the plywood floor, a recurring scene from a play that traded actors for nervous foreign soldiers and family members saturated with hate. The Afghan commando with Deklan began asking the futile questions surrounding the whereabouts of Kasur Halal.

"Eyes down! Eyes down!" The Afghan commandos reminded the father in his own Pashto language. The proud man was reluctant to passively comply with the "eyes down" order, but did so, nonetheless. The Pashto language was interrupted with English from the man of the house.

"May I speak?" The question directed at Deklan, came as semi-clear English from the Afghan villager with his family huddled around him. The Afghan accent was thick and heavy. The sentences were difficult for Deklan and his platoon to decipher but the command of the English language was extensive however hard the words were to absorb.

"Of course." Deklan replied and looked at Josh Needham, a Navy SEAL from Long Beach, California. The two Navy SEALs were surprised to hear the man speak English. They were able to ask the questions and were not forced to rely on their Afghan commando allies to butcher or sabotage the questions and answers they sought. "You can start by telling us your name and then telling us where Kasur Halal is hiding."

"My name is Abdul-Hadi Akbar." The man responded proudly. "My family and I do not have any idea where the man you seek is hiding, nor have we seen him ever in this village. I know about the man you seek but he is not here and has not been here, ever."

"We know he was here. He may not have been in this house, but Halal was in this village. We know he was in the area. You would do your family a great service by telling us where he is." Deklan was direct.

"Are you going to kill us if we cannot tell you where this man is located?" Abdul-Akbar looked at Deklan Novak and for a moment no one else was in the room.

"Would that change your answer if I said yes?" Deklan asked quickly.

"That is not an answer to the question I asked you." The man spoke his brand of Pashto English and at another time and another place,

CHAPTER TWELVE

Deklan would have queried his host about where his language skills were acquired. "I must assume you are going to kill us. I'll try to save my family. I have no other choice. Fighting you and your men would result in the same outcome. Am I not, correct?"

"My brother, we got no time for this." Josh was reminding Deklan to wrap this up.

"We are not here to kill you Mr. Akbar." Deklan was aware of the time restrictions but was not ready to abandon the English-speaking Afghan father. Deklan believed that in some cases when civilian Afghan men were killed as suspected threats, the results had a long-term negative impact for the United States. The children of these slain civilians, "collateral war damage" grew up with hatred for the Western Alliance responsible for their father's death. One civilian death often saw the children of the slain villager, volunteer to join the jihad against the West. The "collateral war damage" was breeding terrorists. Understanding the civilians, Deklan believed, provided a potential path to create less Mujahideen.

"If you have any information regarding the whereabouts of Kasur Halal, then I can make it worth your while to provide us that information." Deklan announced and reached into his tactical utility belt. Navy SEALs were always prepared to do whatever was necessary to complete a task, including bribing the civilian population. "Help us locate Halal and you are helping your family tremendously. If you are hiding Halal or withholding information about his location, that is not going to help your family."

"I simply do not know where the man is, nor have I ever known where this man is located or where he resides." Abdul-Hadi Akbar replied politely and calmly. "Why are you here, sir?" The question surprised Deklan.

"I told you." Deklan snapped. "We are looking for Kasur Halal. He is a Taliban leader responsible for the deaths of many American soldiers and Afghan police officers."

"Deklan." Josh was heading towards the door of the structure. Deklan's SEAL teammate was anxious to exit the village because the information they were looking for was not coming from this family.

They found nothing in the house to lead them to believe Mr. Akbar was harboring the Taliban, so it was time to leave. The other SEALs and Afghan commandos in the room began to leave the home.

"I asked you, why are you here, in my country?" Abdul-Hadi Akbar sought an answer before his intruders departed. At that moment Deklan and the captive family were alone in the room. Deklan assumed Abdul-Hadi Akbar was the only member of his family that was able to speak English, so the conversation was between the two of them. Deklan would join his platoon in short order.

"We are in your country because of the attacks initiated inside the United States on September 11, 2001. Almost 3,000 Americans were murdered on that day." Deklan replied as if he was reading a cue card.

"Again, I see no connection to the attacks and the invasion of my country." Akbar continued. "You stand there in your Kevlar protection, armed with more fire power than most Afghan battalions possess, and you are the blind direction of American retribution. Do you know that more than 60,000 Afghan civilians have been killed in my country since 2001 by American soldiers? These people did not attack your country." Akbar spoke as if he had been rehearsing the speech for a theatrical audition.

"The people who attacked the United States have been trained here and are hiding in the mountains and villages where we search for them." Deklan engaged the premise. "I am here to find the jihadists and the militant Islamic regime behind the attack on my nation. I am here to protect my brothers."

"You are wrong." Akbar announced. "Protection occurs inside your homeland not as an invading force. The United States has spent years trying to find Bin Laden because he is an international terrorist. OBL is in and out of more countries than a U.S. Secretary of State. You are fighting Al-Qaeda not the Taliban. The Taliban is a nationalist organization. They seek power here in Afghanistan. The Taliban seeks to restore order to the Pashtun tribal way of life. If you are here to liberate Afghanistan and help us become a democratic society, we do not wish to become a Democratic society. Either the war in Afghanistan is to avenge the 9/11 attacks or the war is the extension

CHAPTER TWELVE

of American imperialism and the need to control the political process in the region. You cannot have both. Why does the retribution for the 9/11 attacks leave dead children in our villages, where the victims and their families had never even heard of New York City? Why are human rights activists constantly protesting the United States coalition night raids as barbaric and one of the contributing factors to so many civilian deaths in our country?"

"Are you that naïve?" Deklan asked as he gathered up his gear, making certain nothing was left by his platoon. "The Joint Prioritized Effects List (JPEL) is a combined effort between the coalition forces and the government of Afghanistan, your government."

"I am not naïve." Akbar remarked. "But my people are completely naïve to the reason your country continues to assassinate civilians in Afghanistan. The Afghanistan government is financed, supported, and protected by the United States. President Karzai and the Afghanistan High Peace Council hardly can be expected to reflect the wishes of Pashtun society. I know you are a soldier in the United States Navy. You are part of an elite group of highly trained special forces. I know a lot about the Navy SEALs. You are accomplishing nothing. Americans are here to dictate the political process and kill their perceived enemies. Fighting for freedom and the sovereign rights of the United States, you are extinguishing the very foundation you defend. This is our country, not yours. The Taliban are defending their homeland from invasion. They have been defending their homeland for decades. Leave us alone and there will be no more American casualties in Afghanistan. I don't know where Kasur Halal is hiding, but your Special Forces will eventually find him, and they will kill everyone with him. I am sorry I cannot help you today."

"What are the names of your children?" Deklan asked as he gazed down at the three huddled children trying to hide behind their father.

"Ghazan is my son. Gabina and Helai are my daughters." Abdul-Hadi Akbar replied while pulling his kids closer.

"Do you have any other children Mr. Akbar?" Deklan asked. "Have you lost any children to the infidel invasion of your country?"

"I have not." Akbar answered defiantly. "Not yet."

"Excuse the men and I because we have to leave at this time." Deklan took off his helmet and ran his hand through his matted hair. "I can appreciate the level of disdain you speak of concerning the American presence in your country. Can you explain to me how the men defending your homeland, as you have described, can be allowed to take young boys as "chai boys" and excuse the Pashtun practice of raping and sodomizing young boys as an institutionalized practice in rural society. Please correct me Mr. Akbar if I speak lies, but isn't the resurgence of Bacha Bazi now accepted again and not viewed as a criminal act in Afghanistan?" Novak paused and had little time to educate anyone.

"All religion, theology and divinity are man's experiment to explain the inexplicable. Killing in the name of God is explained in the bible in Exodus and Genesis that people are made in the image of God and are allowed to take lives righteously The Qur'an states that God commanded the use of jihad to expand the territory of Islam. Ecclesiastes 3: 1-8 (NCV) says there is a time to kill and a time to heal...a time to destroy and a time to build. Why man justifies territorial genocide as God's command from a manmade ledger will not serve as the inexplicable. Man's nature revolves around justifiable killing as evidenced by every historical period in mankind. That is not only explicable, but also predictable. The divinity in goodness is the inexplicable, which further explains Islamic jihad." Novak heard his platoon mates from outside the structure insisting that Deklan get his ass in gear.

"Your nation does not allow women to work or go to school. Mr. Akbar, the notion that women are for children and boys are for pleasure is not the credo of a passionate and caring society. Taking young boys into servitude is not a symbol of power and status as practiced by Islamic rationale. The existence of Bacha Bazi has glamorized and normalized kidnapping and rape of young boys under the religious arc of sanctioned Afghan pedophilia. Pashtun men come out of the closet and their social status is revered. You'll have to excuse my unwillingness to absorb the lecture you provide under the blanket of pedophilia. I'll also pass on the genocide lessons from the Qur'an. Thank you for the input, though." Deklan slapped his helmet back

CHAPTER TWELVE

onto his head. Deklan Novak dropped a couple 1,000 Afghan Afghani banknotes on a small table near the main entrance to the residence. The brown bills were worth around $30 in American dollars. Abdul-Hadi Akbar looked up at Novak as the Navy SEAL was leaving the mud home with contempt for the emasculating gesture of leaving money. Akbar said nothing. The poverty line for rural Afghanistan in 2009 was an income of $1.25 per day. More than 40% of the rural Afghan population lived below the poverty line in 2009.

The platoon abruptly left the home and the village they had visited less than two hours earlier. Where did Akbar learn to speak English in rural Afghanistan? In a country where more than 57% of males were illiterate, how did Deklan manage to find an intellectual debate that exceeded the daily barracks banter at Salerno? Why were American and NATO forces fighting and dying to defend thousands of proud pedophiles revered in a society that practiced pedophilia more per capita than any place on the planet? The daylight was still an hour away, so the night sky protected the retreat to the base. The split platoon proved not to be a fatal mistake during the night. Corporal Patrick Daniel Tillman may have been looking out for some of his comrades from above in October 2009.

Deklan was never fully vested in the war. Deklan had heard the patriotic calling emanating from the ashes of the Twin Towers, but deployments painted a harsher reality. The civilian carnage surrounding each deployment was not only the collateral damage of a flawed objective, but the civilian losses had also become a non-denominational genocide amplified by combat awards based on kills. The mujahideen leadership in rural Afghanistan and the practice of Bacha Bazi placed a level of validation surrounding the morality of the coalition invasion that transcended Washington D.C. Deklan could survive in the chaos of the directionless chain of command because of the bond that every combat deployment created. The Art of Manliness was formally adopted in 2005, but the Navy SEAL creed transcended the military and became a way of life every SEAL was directed to live by.

. . .

Brave men have fought and died building the proud tradition and feared reputation that I am bound to uphold. In the worst conditions, the legacy of my teammates steadies my resolve and silently guides my every deed. I will not fail.

CHAPTER THIRTEEN

2006-2016...DIVORCE, DARKNESS, AND THE HONORABLE DEATH

GOING into Columbus in late October, 2016, Northwestern was ready to stamp their place on ending the conference doormat moniker once and for all. It had been ten years since Adams took the helm, and everything about the school's football culture had changed at Northwestern except the Big Two mythology. Ohio State and Michigan had to be defeated at least once for a team to be elevated in the conference to be considered a legitimate contender. Coach Adams had made recruiting events newsworthy. Lance Adams had changed the narrative surrounding recruiting, where 4 to 5-star high school football players wanted to play for Lance Adams. The recruiting on-campus visits for the Wildcats became a relevant measuring marker in the conference that brought Evanston, Illinois into the Gameday Power Five college football site selection process for the SEC, PAC 12, ACC, Big 12, and Big Ten conversation. ESPN's College Gameday came to Evanston on October 5, 2013, for the game that pitted #4 Ohio State against #16 Northwestern. OSU won the game 40-30. Northwestern University had re-entered the Big Ten with a splash since Lance Adams had taken the reins. Beating Ohio State or Michigan was the next step to legitimacy. Lance Adams had proven his metal before the OSU game, but the team knew that beating Ohio State in Columbus was key to establishing Northwestern as a football team to be worried about.

In 2016, quarterback Trevor Dunn wasn't sharp but did enough to keep the Buckeyes worried. Dunn was 22-42 for 246 yards, one touchdown and an interception on a tipped ball. Dunn ran for 64 yards and a touchdown. Northwestern was 8-16 on third downs. The game was in doubt inside "The Horseshoe", but the final score was 24-20. OSU won the game. Deklan played much of the game on defense and on special teams. Deklan Novak thought moral victories were reserved for news media pundits and teams that accepted defeat.

Deklan had been struggling within a marriage that began when Amber and Deklan were hammered. Hindsight told Deklan the relationship was doomed from the beginning. Deklan and Amber Bale were both hard charging and wild. The free-fall had started before Deklan joined the Navy. Deklan was partying hard at a local tavern while celebrating a recent MMA victory in Minneapolis, MN when the two strangers bumped into each other while ordering drinks at the bar. The 9/11 attacks had already sparked the Navy SEAL aspirations inside Novak. Amber Bale was 21 years old, and Deklan had just turned 23. Deklan and Amber drank and sang and danced the evening away on the night they met. At the end of the night the only thing Deklan remembered clearly was how his best friend Brian was staring into his eyes, how he pleaded with Deklan to skip the impending romantic interlude. Brian, substantially more sober than Deklan, tried to explain the moguls associated with an intoxicated intimate encounter and the complicated exit strategy that inevitably followed. Obviously, an MMA professional fighter had zero inclination to discard such beauty, and Deklan fell hard.

Deklan Novak smiled, laughed, and assured his MMA teammate that he knew what he was doing. Amber came back to the hotel with Novak that night. They rolled around for a few hours and the future Navy SEAL was hooked. Amber clearly provided the right amount of chaos, pain, and comfort in Deklan's life. Deklan now had a partner to ride the roller coaster together. It was early 2006. Within weeks

CHAPTER THIRTEEN

Amber moved into Deklan's place, and they spent all their free time together. It was wonderful at first, then became increasingly challenging and confrontational.

Over the first year, Deklan attempted to break up with Amber several times but could not follow through. The more time Deklan spent away from Amber. the more he needed her. When Deklan finally left for Navy boot camp he was thrilled to be artificially cut off from Amber and the chaos they shared for a guaranteed nine weeks of solitude. Yet, two days before boot camp graduation at Great Lakes, Deklan Novak proposed marriage and Amber Bale concurred. Amber agreed to accompany Deklan to San Diego for BUD/S training where they would be married at a California courthouse a few weeks after Deklan had completed Hell Week. Deklan recalled years later how chaotic and reckless and self-destructive he had been prior to the marriage proposal. The sense of invincibility that emerged in some SEAL candidates was empowering and at the same time, it was well beyond dangerous. The complexity of an alcoholic in love inspired the Navy SEAL to wonder how he was still alive. What had been left of sexual compatibility and physical intimacy passed through a bottomless silence that had once defined language through pleasure, but now ended each night with a lack of interest coupled with unfamiliar silence.

The years of marriage saw great personal gain for Deklan that came as a solo artist. BUD/S, SEAL Team One, deployments, a prestigious college diploma and a stint playing Division 1 college football in his 30's, brought personal accolades and marital friction. The summer of 2016 came with the faded image of a Navy SEAL Trident inscribed in Deklan's head and a Sig Sauer P226 barrel pressed against the top of Deklan's mouth. The thought of leaving four boys alone in a chaotic world with their bi-polar, alcoholic mother was the only thing that kept Deklan alive on that day. Serendipity rarely emerged in Deklan's marriage, but the final curtain image continued to hover above the desire to raise children in the chaotic whiskey fog that Amber and Deklan lived inside of. The hovering whiskey cloud was not an entirely accurate picture of the truth. In some distant reality,

Deklan's four young boys and the lives they had before them, were not enough positive energy to derail the destruction on the track that afternoon. The engine was running, and the alignment held firm. Fortunately, there was another force that came roaring into play that afternoon.

It was a beautiful sunny summer weekend afternoon in Northern Illinois. Deklan had three hours of an Absolute vodka buzz well nursed, and he had just driven off-road with his perfectly clean Glacier Blue 2015 Ford Raptor pickup truck to the location he had picked out long ago. Deklan had half of a bottle of Absolute remaining and with the truck in gear, the clean vehicle would be a perfect ride for the grand finale. Deklan cranked the volume high and played raucously over the factory speakers "The Kiss." The song was the theme song from the opening scene of **The Last of The Mohicans**, *a 1992 film*, that Deklan played on repeat and vowed that he would finish the bottle of vodka before playing out his final solution. Deklan and his brother DJ watched **The Last of The Mohicans** a dozen times along with **Legends of the Fall and Heat**. Brad Pitt as the mysterious Tristan in **Legends of the Fall**, became an indelible iconic loner that Deklan connected with from day one. From **Heat**, Deklan saw Neal McCauley in the mirror. Hawkeye's confrontation with Magua in **The Last of the Mohicans** was honor in death. After DJ's death, Deklan spent many nights listening to that song and watching those movies imagining a life where DJ did not die.

Summer camp was not far off, but Deklan dreamed of a life where he had not abandoned his brother in purgatory to pursue becoming a Navy SEAL. Deklan imagined that DJ would still be alive if in their final conversation, Deklan had not told his brother what a "fucking weak piece of shit" he was for relapsing back into his heroin addiction again. Deklan pitied DJ but pitied himself much more without ever admitting that a Navy SEAL was that vulnerable. Deklan finally pitied himself from the depression that was consuming him, and he pitied everyone around him. What had all the accomplishments meant? Why was the depression so much stronger than he was? That was impossible. Life had gotten very dark, and there was apparently only one way out.

As the bottle neared its end, so did the clear blue sky and the fine late summer day. A literal darkness rolled across the sky like a wet wool blanket as Deklan swallowed another rancid layer of warm vodka. As if the blanket were being wrung out by God-sized hands, the sky began releasing raindrops the size of basketballs. The first lightning strike put the storm 30 minutes away but by the second lighting strike, the storm was seemingly already directly overhead. The rain started coming down in sheets, hard and damaging in an unusual lethal torrent. Deklan stared at the massive drops as they dispersed the dry clay soil upon impact like miniature asteroids slapping against the loose dust of the moon. Several small streams of fresh clear rainwater began to accumulate next to the driver's side front wheel and Deklan marveled at how quickly the rain pellets converged and gathered to become a deluging slurry of wet mud and clay. Lightning preceded a crush of thunder, and the atmospheric explosions grew steadily closer together until there seemed to be no perceivable silence between strikes. Deklan thought of the times as a kid he would grab his .22 caliber pistol during storms so he could fire two or three blank rounds out from his back patio door into the exposed tree line.

Under the cover of the next thunder blast and now as an adult consumed with an Absolute vodka visual lifeline, Deklan Novak pushed his Sig Sauer P226 out past the Ford Raptor door jam and into the storm directing two shots of 9 mm rounds into the wet clay below. Then came another violent reply from God along with an electrical lightening display like an environmental fire hazard connecting with the parched terrain. Novak's retort to the Almighty was to fire three more rounds into the dirt. Again, God fired back and so did the intoxicated Navy SEAL. Deklan yelled into the wet darkness while God and the Breacher fired back at each other repeatedly. Lt. Dan's tirade high up on the mast in **Forest Gump** ran through Deklan's intoxicated mind…" You call that a storm? I'm right here, come and get me." Then the pistol went click. Click. Click. The magazine was empty. Deklan had only brought one magazine, never imagining that the mission would exceed 17 shots to complete. Deklan was Winchester, out of ammunition. He was drunk, wet, and cold. And now the truck was stuck in a torrent of wet clay. But Deklan was alive. He smiled and laughed and stared long into the eye of that storm. Deklan had wrestled with God and was still alive. The veteran Navy SEAL shook his head in awe. Defeated, yet content and soaked to the bone, Deklan proceeded to walk home, maybe four to five miles to the ranch house he and Amber owned. That night Deklan vowed to never touch another drop of alcohol, although he had hoped his memory would fail the next day and sobriety would be, as usual, a path to follow for someone weaker.

A few weeks after the storm and the thunder vs 9 mm bout with God, Amber and Deklan separated for the final time. On October 29, 2016, on the way home from Evanston after the Northwestern vs. Ohio State football game in Columbus, Deklan and Amber got into a major league fight in the truck following the team's arrival at O'Hare and the bus ride to Evanston from the airport. Deklan exited the vehicle at a stoplight and ordered an Uber back to campus. Deklan bunked at a teammate's apartment. Amber returned to Kenilworth where the boys were staying with her parents.

Life after divorce was as hard as anything Deklan had ever done in his life. Why did it hurt so much? Why did he feel so empty? Why did he feel like he could not live without her? Why did he feel like he was dying inside? Deklan had no answers to those questions. He had spent his life clinging to things that made him feel safe and strong and repelling or running from all the things that did not. Deklan Novak spent his life moving to and from safe harbors managing his fears. Deklan simply reacted to the way that he felt rather than understanding why he felt a particular way. Novak had always assumed that was how everyone moved through life. As far he could tell that was how his parents operated, and their parents operated, and he presumed their parents operated in the same manner as well. The dysfunctional father that defined his childhood must have been buried in the cerebral cortex where higher thinking was supposed to prevail. Deklan eventually could not shake the inexplicable progres-

CHAPTER THIRTEEN

sion of knowing that his marriage to Amber was not how men and women were designed to move through life successfully. Finally, through a haze of vodka or a gun battle with no one, Deklan Novak refused to run away from the pain. Nor would he run to the things that once brought him escape and comfort. Deklan knew that he had to understand the pain before the battle to overcome the pain could be waged.

For two years, Deklan did not sleep or eat well. After football ended in 2016, Deklan's body went from 240 lbs. to 205 lbs. The nights were long and awkward. The Navy SEAL sat up awake most of the night, often in tears, angry, confused, scared, and humiliated. Deklan was in constant pain that he did not admit or understand. The unknown was terrifying, especially for a man trained to be a machine, incapable of being terrified.

The small moments of clarity never came, and Deklan assumed his life had to add up to one thing. Deklan dedicated his life to his four boys, assuming God's lack of direction meant the direction was crystal clear and God was maybe subtle but brilliant. The boys and their well-being were all that God wanted Deklan to be concerned with. Divine guidance only became cloudy when the search was buried in self-wallowing narcissism. Ultimately machismo was a primary form of insecurity that Coronado, the octagon, and the public perception of the SEAL brand wouldn't reveal until Deklan was finally able to let go of the shield temporarily. Deklan quit his job, as a lucrative weapons sales' representative selling arms for Northrop Grumman, a $36 billion American defense contractor that sold arms and technology mainly to the United States Department of Defense and the C.I.A. Deklan stopped traveling. Deklan moved permanently to the 53-acre ranch in Wisconsin. Deklan sold everything he could and learned to live smaller and more efficiently. Deklan learned balance, humility, and silence, working sometimes for days alone in nature. He learned the ways of the animals, of the Earth, of the Universe, of energy. Deklan Novak re-engaged with his brother, DJ, and what DJ tried to teach Deklan when he was alive. Deklan was trained in anger and DJ understood human nature.

Deklan began to understand that his anger wasn't about Amber. Divorce was just the striking point. Sobriety, divorce, the military, and generational PTSD all converged, and the bundle needed to be unpacked. Yet at the precipice of convergence, Deklan knew intuitively that somehow there must be some better way. Deklan also knew intuitively that he could no longer avoid any of the recurring feelings that had defined his past years of turmoil. The summit was an elusive peace that would take a few more years and the FBI to locate.

Deklan followed every lead with SEAL-like scrutiny. He attended a thousand AA meetings, countless hours of therapy, a thousand hours of meditation and breath work. There were hundreds of hours of prayer, cold plunges, Shamanism, fasting of every kind, celibacy, psilocybin trips, hypnotism, regressions of many types, and Deklan read hundreds of books covering every field of healing and self-inquiry. Deklan kept dozens of written journals, attended Christian retreats, Agnostic retreats, Buddhist retreats, and Pagan retreats. Deklan became a bird dog seeking out any modality claiming a special access pass on the inner journey. Novak squeezed every drop from every source. Nothing was above or below the veteran SEAL. In the process of rediscovery, Deklan gradually grew to love himself a little more each day. Then half-way through his third year of sobriety as a single father of four boys, the miracle happened when Deklan Novak realized that everything was going to be ok. He realized that he was ok. That nothing was ever out of place and that he was exactly where he was supposed to be. Deklan realized that everyone was exactly where they were supposed to be. Deklan found peace and felt peace. He came to know peace. Deklan reconnected with the SEAL Trident he had metaphorically placed in storage. In exchange for all the pain, Deklan was granted peace. And in the process, the former SEAL Team One breacher acquired and developed with absolute SEAL-like proficiency all the auspicious tools needed to maintain the peace he had found.

CHAPTER
FOURTEEN

NOVEMBER 15, 2014...NOTRE DAME FIGHTING IRISH VS NORTHWESTERN UNIVERSITY, 2:30 P.M. KICKOFF

THE NOTRE DAME game was out of this world. Deklan always found it difficult to condense down into sentences and paragraphs with any sincerity and accuracy what playing football at Notre Dame Stadium was like. The grandeur of the endless facets of moving parts taking place all around Deklan, the NU football team and the stadium resting under the watchful eyes from a mural on the south panel of the Notre Dame library tower called "Touchdown Jesus" was for the most part indescribable. How does one describe what is only alive within the senses of those allowed to participate.

Notre Dame Stadium was built in 1930 under the guidance of Knute Rockne, regarded as one of the greatest coaches in college football history. Rockne's legacy spawned the nickname, "The House that Rockne Built." The stadium seating capacity was just under 60,000 fans. In 2014, a $400 million renovation project was completed and increased the seating capacity to nearly 80,000 fans. A statue of Knute Rockne stood near the north gate of the stadium.

Northwestern shocked the college football world in 1995 by

finishing 10-1 and earned a berth in the Rose Bowl. The last time Northwestern and Notre Dame met was the season opener on September 2, 1995, in South Bend on the gridiron of one of the most iconic and recognizable collegiate stadiums in the nation. Coach Lance Adams played middle linebacker for the Northwestern Wildcats that Saturday afternoon in a Wildcats 17-15 victory over Notre Dame. Adams recorded eleven tackles in Northwestern's victory over #9 Notre Dame in South Bend, the Wildcats' first victory over the Irish since 1962. That season against #7 Michigan, Adams led the defensive effort with 14 tackles, including two tackles for losses, in the Wildcats' 19–13 win, the first for Northwestern against Michigan since 1959. At one point during the 1995 season Adams averaged over 13 tackles a game on his way to Consensus All-America honors. Lance Adams was unable to play in the Rose Bowl after breaking his leg in the next-to-last game of the 1995 season against Iowa. USC defeated Northwestern in the 1996 Rose Bowl 41-32. Adams returned for the 1996 season, leading the Wildcats to a 9–3 overall record, a second straight Big Ten Championship, and a second consecutive New Year's Day bowl, the 1997 Citrus Bowl.

Nineteen years later in November of 2014, Notre Dame and Northwestern met again. And again, another game was destined to be remembered as one of the greatest games in Northwestern football history. Coach Adams and the coaching staff at NU made it very clear that there was no looking ahead of this week's game to later perhaps more "important" games in the season. "The season is now! The season is this week!" Adams was adamant without cliché rhetoric. The game was on. The Notre Dame game was a true exception. At least as far as Deklan could tell. While the team started off each new game week with the same focus and determination and intensity, Novak could tell that the Notre Dame game had a special meaning for Lance Adams. Deklan also knew he was on a cloud that didn't appear to be dissipating any time soon. Deklan played high school football, watched Division 1 football religiously and had dreams, but never expected to be a starting middle linebacker for a D-1 team heading to the cathedral of football…the House that Rockne Built.

CHAPTER FOURTEEN

Any number of Notre Dame traditions could have made the list of recognizable/iconic college football traditions. There were the football team's magnificent gold helmets, made with fragments of real gold. There was Touchdown Jesus. There was Rudy. The most iconic tradition, not the oldest tradition, was the locker room sign, "Play Like a Champion," that players whacked before making their way into the tunnel at Notre Dame Stadium. The saying and the sign were originally coined in the late 40's by Bud Wilkinson with the Sooners at the University of Oklahoma, then brought to Notre Dame by legendary head coach Lou Holtz in 1986. There's a scene in the movie, "Rudy" with the Lou Holtz sign. Rudy played in 1975, making the cinematic reference to the sign historically inaccurate.

Friday afternoon the team left Ryan Field in Evanston at 3 pm for the hour-and-a-half drive to Michigan City, Indiana aboard four deluxe charter buses. After off-loading the personal gear and getting checked into their rooms, Deklan and his teammates began meeting as a team, as position groups, and with the medical staff if needed. The team then met in the banquet hall for dinner at 6:30 p.m. and the rest of the night was spent as guys trickled in and out of the media room upstairs, where everyone watched Tom Hardy's "Warrior" on a wall-to-wall projection television screen. Novak loved that movie so when he was asked during the leadership council what he thought would be the most appropriate film to watch in preparation for Notre Dame, it was either "Warrior" or "300". Northwestern University was going to need other-worldly gritty individual performances from every guy on the squad as a team that rolled into enemy territory heavily outgunned and outnumbered. Notre Dame was ranked 15th in the A.P. poll and the Wildcats had just lost four straight games.

When the team arrived in South Bend aboard the four luxurious charter buses Saturday morning and exited the chariots in full suit and tie as the team did before every game, the mood held a tangible level of tension. In the guest locker room, Novak quickly changed into a pair of joggers, a NU hoodie and purple Under Armor gym shoes. Deklan immediately walked out of the north tunnel across the end zone and out onto the field. Deklan jogged to the 50-yard line and pulled in several full lungs of crisp November Indiana air. The Navy SEAL took in the glorious sight of workers, and staff scurrying around a mostly empty Notre Dame stadium while they carried things to prepare the stadium for the 77,000 plus fans that would attend the day's sold-out event. Novak ran the entire length of the field three times at full tilt. Deklan ran around the outside of the field several times, then made his way back to the locker room to get suited up for team pre-game warmups.

The locker room was a dingy tiny little enclave that had probably not been improved since Knute Rockne himself designed and had it built in 1930. The lockers looked like they were salvaged from the Catholic middle school down the street that was demolished a decade earlier. The lockers were painted a drab brownish yellow that would depress Thomas Aquinas. The rest of the room was covered in scuffed up white paint and the floor was loosely covered by an ancient green industrial carpeting that had seen much better days. The carpet smelled like an overused YMCA sauna.

None of the aesthetics mattered to Novak nor to any of his teammates. Deklan, like the rest of the roster, got suited up quickly, and Deklan made his way to the training room to tape his ankles and left wrist that was injured the previous week in the loss against Michigan. One hour before game time, everyone made their way to the field to get warmed up in position groups and as a team. The stadium was filled about one third capacity at the time, yet the energy was palpable and electric, a small precursor to the afternoon that was going to be. The pregame voltage felt like the Superbowl to Novak. Fans continued to trickle in as the team moved through the well scripted pregame schedule. Voices from the stands would occasionally yell one of the player's names or numbers. Family members and friends coagulated in the first row at the 50-yard line attempting to steal a moment from their friends or loved one's time before the team ran back into the locker room. A few of Deklan's family members and friends had worked their way to the bottom row and called for his attention. Deklan ran over and shook a few hands then followed the

CHAPTER FOURTEEN

team back down into the bowels of Notre Dame Stadium, like the Spaniard in Gladiator played by Russell Crowe in the 2000 film. On the most important football Saturdays of their lives, Lance Adams spoke to the team before they left for the field at Notre Dame Stadium on November 15, 2014:

"We will play for a national television audience tonight. We will play before 80,000 hostile fans tonight. You have sacrificed much to make it to the highest level of college football. Tonight is not the night to watch it. I expect each player on our team to be engaged with every play and every movement on the field. I expect every player on our team to play each down like it might be your last down as a college football player because it might be. I'll replace any player tonight that doesn't act like he may never see the field again. You can keep the scholarship, but you won't see another day in uniform if you don't compete like an injured pit-bull in a dog fight on every single down no matter what side of the ball you are on. Games are never won or lost by luck or chance. Games are always won by teams that play every down relentlessly because losing is not only unacceptable, losing is embarrassing. Tonight, we will not be embarrassed by a Brian Kelly team. Notre Dame is a national treasure and according to an article published in Bleacher Report on June 1, 2011, Notre Dame is America's team not the Dallas Cowboys. The article argues that Notre Dame has a longer history than the Dallas Cowboys, has won more titles, and has played in more games hailed as the "game of the century." The coaching etiquette manual provides guidelines on specific public assertions directed to individuals. I never read the manual. Brian Kelly became the Head Coach at Notre Dame in 2010, and I personally would appreciate your help to send his ass back to Cincinnati. My apologies to any faith-based denomination that I may offend here today, but we will not take a knee and pray before the game or during the game. Chaplains belong on the battlefield or looking for God on Death Row. Chaplains do not belong in a Big Ten locker room to pray for us to kick the crap out of Notre Dame or to look for heavenly assistance on fourth quarter drives that need Al Pacino recalling the "margin of error" speech much more than we need a confes-

sional. Say grace before a meal at the dorm. Kiss the crucifix around your neck or the Star of David and tuck it back down into your jersey. Leave the football field to me. We'll take a knee and pray after the game that the team we just kicked the shit out of can make it home in one piece. We'll take a knee and say a prayer for an Irish kid from the Chicago area. (Many of the NU players were confused about the Irish kid reference but Deklan Novak knew about the accident. Coach Adams did not elaborate further.) *Game-time people will be here in a few minutes and that will be the wake-up call for what you want the entire nation to say about the Northwestern Wildcat football team! Take in the breathtaking silhouette as we exit the tunnel at Notre Dame Stadium and take the field tonight, then prepare to tear that shithole down."*

Notre Dame football was the gold standard for college football in the United States. As much as the nation hated to have the first NCAA national championship four-team playoff picture dominated by the Fighting Irish and the Crimson Tide, the facts were undeniable. Deklan cared about gameday only. The Wildcats had the top ranked Big Ten defense in yards allowed per game. The Wildcats won and lost record for the season did not reflect the offensive and defensive statistics and where they ranked in the conference based on those statistics. The Wildcat defense allowed 241.6 yards per game in 2014 after 9 games. The NU defense was fourteenth nationally in points allowed per game at 19.7 and the offense was ranked sixth in the Big Ten in points scored per game, a remarkable statistic for a team riding a four-game losing streak. Deklan Novak started the last seven games in 2014 as the Mike linebacker for the Wildcats. Novak led the team with 12 sacks, 118 solo tackles and was second on the team with three interceptions.

Head Coach Lance Adams paced around the visitor's locker room as the team gathered around him. The game was about to begin. NBC had the national telecast as part of their television contract with Notre Dame football that ran through 2025. Tom Hammond and Doug Flutie called the game on November 15, 2014. Adams had said much of what he wanted to convey before leaving the locker room for

CHAPTER FOURTEEN 159

the walk to the field. Lance Adams looked over his team. The players got quiet. The visiting locker room at Notre Dame Stadium smelled like ammonia and perspiration. Adams stopped and faced his team in front of the lockers. One look from Adams had the players ready for a brawl. Military acronyms were overused in college football and professional football. The players were not heading into battle or going to war.

Two coaches hunkered down and stayed in an office while the players waited in the locker room. Lance Adams lingered with his defensive guru until the team was ready to leave the locker room. Adams turned to Behr Thomas and said, "Notre Dame is a 17-point favorite tonight. Northwestern is 2-14 against the spread over the last two years. My dad put some money on us to win tonight at +800, not just to cover. I don't gamble. Predominately because my job prohibits gambling, but I hate losing money. I take losing personally. In Las Vegas, if I lose money at a black-jack table, I hold the dealer responsible. I want to break his jaw. That's not a good premise for a gambler. I just have a feeling that my pop is going to be a happy camper later tonight. Let's go, coach." Lance Adams smacked his defensive coordinator on the shoulder like players do in pads. Behr Thomas smacked his head coach back hard. The two men left the office prepared and ready.

The next time that Deklan Novak headed towards the tunnel that led to the field, he was carrying the United States of America Flag standing next to Coach Lance Adams. The pair waited at the entrance to the north tunnel with 69 warriors chomping at the bit behind them. In the FBS only 70 of the 100 plus players on the team were allowed to travel for away games. The stadium was filled to the brim and overflowing with wild intensity.

Adams planned it perfectly. Northwestern was supposed to be the second team out of the tunnel. The home team entered the field first. Adams knew how the home team and guest team locker rooms were oriented to the north tunnel which led down to the field. Adams knew that if the Wildcats exited from their locker room before the Irish, the entire team would be standing on the other side of the tunnel staring at each Notre Dame player as they squeezed through the doorway of their locker room towards the field. Coach Adams and Deklan Novak, the Navy SEAL that the Irish had heard so much about, and the entire Wildcat Football Team stared down the surprised Fighting Irish. The Wildcats hooted and hollered, growled and howled, yelled and grunted like a pack of 71 wild pit bulls. The Irish football team looked shell-shocked, and the game had not started yet. After Notre Dame ran out onto the field, Northwestern shuffled down to the threshold of the north tunnel and Coach Adams and Deklan ran the Wildcats out past the north endzone across the 50-yard line and back to the opposing sideline. Brian Kelly was pissed about the tunnel crap pulled by Lance Adams. Brian Kelly was not destined to be a happy man as the evening played out.

The game lived up to all the hype. Notre Dame was up at halftime 27-23. The Wildcats went into halftime knowing they could hang with those guys, knowing that they could beat those guys. Novak became a screaming mad man for much of the first ten minutes of halftime. Then two minutes before returning to the field coach Adams took over. He called the entire team in close. Everyone gathered around the head coach, and he paced the locker room like a hungry lion.

> "I did make a big deal about this game, because you all knew how important it is to this team to this season, to me as a man and a fellow Wildcat." Lance Adams spoke from his heart. "And every one of you put in the time, put in the effort to be here, to be where we are right now within striking distance of coming into Notre Dame's house and punching them in the face repeatedly until they break. Until they say they have had enough, and until they quit. I didn't talk about this much, but I want you all to know that the last time I was in this stadium, the last time I played on this field, the last time I was in this locker room, and it doesn't look any different than it did, was 19 years ago. That was when my Wildcat teammates and I walked into this place as 28-point underdogs. Nobody predicted that we would win that game. Nobody believed in us besides ourselves. We did what everyone told us was not possible. We won. And I want to show my team today what my contribution to the

Irish legacy was. You see this fucking dent right here (as he pointed to a dent at about five and a half feet above the ground on the side of one of the dusty old lockers). Do you see this mother fucker!! I made this dent nineteen years ago when we came out of our halftime break and kicked the living shit out of those guys on their own damn stadium turf." Lance Adams began to pound his fist into the same dented locker as he did nineteen years before, crushing the locker further and shaking the entire row of connected lockers with it. "This is what we are going to do to them right now! Let's finish this fucking team now!"

Coach Adams looked and sounded like Herb Brooks (Kurt Russell) from **Miracle** where the coach admonished his players to grab the moment at hand, as "great moments were born from great opportunities." The entire team erupted in carnality. The team tore that room apart like Deklan Novak had detonated a string of large satchel C-4 charges around the perimeter of the locker room.

After three quarters Notre Dame led 34-26. The Wildcats exploded at the start of the fourth quarter with a 91-yard drive for a touchdown. Notre Dame led 34-33 with nearly ten minutes left in the game. Punts were exchanged. The Wildcats had burned a time-out early in the second half by sending too many men onto the field. The Wildcats were driving towards Notre Dame territory as the fourth quarter progressed and time was a growing factor in the game. NU wide receiver Ronnie Berry ran deep on a post pattern that had not been called by the sideline. Coach Adams grimaced because he knew by the defensive Irish alignment prior to the snap, that Wildcat quarterback Trevor Dunn was going to check the play down from a draw play to combat the Irish eight-man front.

"Four Whiskey! Four Whiskey! Blue 31, Blue 31, hut, hut!" Trevor Dunn scanned the field and barked out from a shotgun formation.

"Four Whiskey" changed the play to a post and announced what receivers were the 1st read, 2nd read, and 3rd targets once the ball was snapped. "Blue 31" identified the defense. Berry was as fast as a cheetah, and he left the line of scrimmage like he was shot out of a cannon. Berry caught the ball in stride and was headed for the go-ahead score.

A Notre Dame safety had taken the correct angle to catch Berry. Ronnie Berry met a stone wall called Xavier Dowler, a two-hundred-and-ten-pound sledgehammer DB with cleats. Berry was annihilated and fumbled the ball into the waiting arms of the Fighting Irish. The Irish immediately turned the play into six points, missed the extra point and led 40-33 with less than five minutes left in the game.

The teams exchanged punts again and the clock was not on the Wildcat's side. With less than three minutes to play in the game, Notre Dame had the ball on the NU 29-yard line with a third and ten. Another score and the game was effectively over. Irish quarterback David Motor (not a better name possible for a QB), was in shotgun position. The entire stadium assumed that the Irish would run the ball to the center of the hash marks, hope to gain the ten yards for a first down, but settle for a field goal, that would effectively end the game. Motor took the snap and veered to his right with a play-action fake. David Motor pulled the ball back from the running back and set up for a pass to the tight end breaking for the corner of the end zone. Deklan Novak had spy assignment on David Motor. The second the ball was snapped; Novak knew that Motor was not going to run the ball. Motor was in a tight battle for the Heisman Trophy and the total touchdown passes for the season could have potentially decided the Heisman vote. Novak broke on the route before the tight end knew who was covering him. The pass came in a half-step behind the tight end but not a bad throw. Deklan was already there and running at full speed up the field when the pass was intercepted at the five-yard line.

Ninety-five yards to go and Deklan had clear sailing to the opposing end zone. The stadium went silent as Deklan ran towards the first touchdown of his college football life. The only hiccup in the plan was Deklan's open field speed. As Chris Berman might have belted out…" there goes Novak, rumbling, stumbling, bumbling to the goal line…he could go all the way." Notre Dame wide receiver Renee Sojurn, a world class track star, ran Deklan down like a lion chasing a water buffalo. Novak was tackled at the nine-yard line of the Irish. NU's quarterback dropped back on the first play for a planned quar-

CHAPTER FOURTEEN

terback draw and waltzed into the end zone untouched. The game was tied 40-40.

Deklan Novak was not the captain on defense, but the team huddled on the sideline where Coach Adams called on the kickoff coverage team to do their jobs. The defense knew the game was in their hands after the kickoff. The defense watched the kick come down in the end zone for a touchback. Deklan grabbed the other two linebackers before they went back out onto the field. There was a television timeout on the field. The rest of the defense gathered around and waited for the Navy SEAL to say something. Deklan's message was delivered over the infamous Irish hometown crowd noise. Deklan was screaming with a combination of saliva and November sweat flying from his facemask:

"We are not losing this fucking game!" Deklan's voice was crystal clear, and his eyes burned pure energy like two stars exploding against the night sky. "This shit is for real now...we are the mother fucking Titans!" Deklan screamed as loud as possible. "They don't gain a fucking yard! You hear me? Not a fucking yard and I'll make sure they fucking remember this platoon for the rest of their fucking lives!"

Deklan's decibel level pierced the air like transcending F-84 twin turboshaft supersonic jet engines that required ear protection. Deklan's tirade carried down the entire Northwestern sideline. Novak and his defensive teammates exploded from the sideline like a herd of Brahma bulls. The time was theirs and the moment was now. Notre Dame ran the ball three times and did not gain an inch. The Irish lost four yards on the first play and the next two plays went for no gain. Adams dared David Motor and the Irish to throw the ball. The Wildcats put eight men on the line of scrimmage. The defense was not a decoy, but the Irish felt that NU would back out of the front each down and that never happened. The cat and mouse game between coaches during those three downs was won by Lance Adams. Brian Kelly was apprehensive about another interception, and Kelly chose to play for the overtime session. The Irish had used their timeouts. The

Wildcats would get the ball back with under two minutes to play. They did not want to go to overtime.

The Wildcats were on the Irish 30 with a first down and 24 seconds to play. Dunn brought the team to the line, and he knew the DB's were lined up to give up a ten-yard pass, but nothing was supposed to get behind them. The nickel back was supposed to roam the middle in case the tight end planned to knock out half the yardage unimpeded. The two outside backers blitzed. The play was a three-step drop and the backers could come free. The ball would be gone before they could reach Dunn. Trevor Dunn took the ball from under center. The tight end leveled the nickel back on a cross pick and Micah Hart, the running back was lined up in the slot to Dunn's left. The outside backers blitzed and ran past the play. Dunn took three short steps back and hit Hart over the middle at the 15-yard line and it was a foot race to the end zone. Micah Hart side stepped one safety and dove for the goal line. Touchdown. The Cats were not in South Bend to leave a shred of doubt about who had been in town.

Lance Adams held up two fingers. The Wildcats were going for a two-point conversion. NU led 46-40. If they failed, the Irish still had a chance to win the game, although highly unlikely with seconds left. Lance Adams was going for two points to make a point. Shotgun formation, empty backfield and Trevor Dunn called for the ball from a deep drop. Once he took the snap, Trevor cradled the ball and ran forward in an apparent quarterback draw once again. Two steps into the run, Dunn stopped on a dime, jumped in the air, and hit the tight end Booker Elliot for two points and a massive statement with 15 seconds left on the clock. Pandemonium erupted on the Wildcat sideline. The score was 48-40. Adams went berserk trying to settle his sideline down. The game was not over yet. Finally, the Wildcats managed to collect themselves and refrain from dumping Gatorade on their head coach before the game had been concluded. 3-2-1…the game was over. The bus ride for all time awaited the Northwestern Wildcats and their tatted-up frogman.

Deklan Novak wandered around the field in ecstasy. Deklan celebrated with the players on the 50 yard-line. Novak greeted his freezing family and friends. They all formed what appeared to be a rugby scrum. Deklan and his family hugged and kissed as Deklan carried his two oldest boys around the field in an abbreviated victory lap. Novak was ecstatic and exhausted. True exhilaration came from triumph. "Class 264 secured, Hell Week." "Northwestern 48 Notre Dame 40." The family took pictures and enjoyed the triumph together.

CHAPTER FIFTEEN

2008...RECEIVING A NAVY SEAL TRIDENT

HOW COULD Deklan have known that his graduation and the week spent in Coronado with DJ would be the last time Deklan would see his brother in person. DJ's timeline was nearing an end while Deklan's new life as a SEAL was just beginning.

Deklan had to work a few hours in the morning. Novak had to make sure everything was set up properly on the grinder for the ceremony on the following day, double check his dress whites and get a fresh haircut before the first round of relatives arrived at 1 pm. Back and forth from the airport all afternoon and by 6 pm everyone was checked into the Coronado Island Marriott Resort and Spa, unpacked, and waiting in the lobby ready to hit the town of Coronado.

Most of the arriving family members and friends had started the day drinking on the plane. It was not difficult to extend the celebration. Deklan and his entourage laughed and talked and cheered. Deklan kept his drinking to three or four beers and tapped out early at 9 p.m.

The night was nowhere near over for the family. Deklan said his final goodbyes for the night and walked out the front door of his buddy JD Christie's bar called Coronado Sports and Spirits. Deklan heard a chant behind him that was beginning to grow. With a ridiculous half-embarrassed smile on his face, Deklan stopped and turned around to see 30 of his family and closest friends chanting "Novak, Novak, Novak" while pumping their hands in the air in rhythm. JD Christie behind the bar joined in and instantly the entire place was yelling Novak's name. Deklan smiled and shook his head. Simultaneously, Deklan wheeled around and threw both hands in the air like Rocky Balboa at the top step in front of the Philadelphia Museum of Art. The soon-to-be official Navy SEAL walked through the front door in a triumphant exit. The place erupted in a thunderous ovation. The raucous support was heard down the next block as Deklan walked home feeling like he had done it. He had conquered life. Deklan Novak had left his old life behind and emerged now as something new. The perfect weekend was about to begin in earnest and Deklan was poised to celebrate and enjoy the victory with his family. Deklan's father had not arrived yet.

The guests began arriving at Naval Amphibious Base Coronado two hours before the ceremony began. This was the only time that the base was open to civilians without a security clearance from a SEAL and an official escort by a SEAL. With an invitation to the Trident ceremony, guests were able to drive through the double front security gate and park just inside the base in the NEX shopping center parking lot. From there they had a semi-short 500-yard walk to the entrance of BUD/S first phase quarterdeck where Deklan and John Cook began their Navy SEAL journey some 18 months earlier.

Coronado is a resort city located in San Diego County, California. The city sits across the San Diego Bay from downtown San Diego. Coronado was founded in the 1880's and is often referred to as a tied island connected to the mainland by a tombolo (a sandy isthmus) called the Silver Strand. Silver Strand Training Complex South (SSTC-S) is the premier training facility for the United States Special Forces. The 578 acres are located between Imperial Beach and Silver Strand State Beach. Coronado is Spanish for "crowned" and therefore the resort city is nicknamed **The Crown City.**

The premier hotel on the island is the Hotel del Coronado, 1800

CHAPTER FIFTEEN

Orange Avenue. Founders Elisha Babcock and Hampton L. Story, along with San Diego developer Alonzo Horton bought the island in 1886 and in 1888, they built a spectacular hotel featuring the unique Queen Ann architecture with wedding cake trim and red roof turrets. The project originally was slated to cost $300,000, but the budget ballooned to over $1,000,000 in the late 19th Century. The result was a 365,000 square foot hotel with 680 guest rooms that still stands today as the face of Coronado. Marilyn Monroe was a frequent guest and in 1958, segments of "Some Like It Hot" were filmed at the iconic hotel. Rooms were booked for $2.50 per day in 1893 and now the average cost to stay at the Hotel del Coronado hovered around $750 per night. The hotel is located near Coronado Beach, some 2.4 miles from the north side of the island where the Coronado Island Marriott Resort and Spa was located. Deklan Novak's family and friends had reservations at the Coronado Island Marriott Resort and Spa.

The fun never ended as prospective Navy SEALs moved from the completion of BUD/S Phases 1, 2, and 3, to their new SEAL Teams. Even numbered teams (SEAL Team 2, 4, 6, 8) were stationed on the East coast in Virginia and classified under the Naval Special Warfare Command. Odd numbered teams (SEAL Team 1, 3, 5, 7) were stationed in Coronado, California under the Naval Special Warfare Group One. Deklan Novak was assigned to SEAL Team One on the west coast. The new guys to the Teams had another four-month mandatory course to pass called SEAL Qualification Training (SQT) before receiving their Tridents. BUD/S was mostly about the physical conditioning necessary to become a Navy SEAL. The SQT course training covered the major skills required to conduct SEAL missions, including but not limited to Hydrographic Reconnaissance, Communications, Field Medicine, Air Skills, Combat Swimmer, Land Warfare, Maritime Operations (long range ocean navigation), and Submarine Lock-in/Lock-out. Deklan heard the motto in SQT a thousand times…the more you sweat in peacetime the less you bleed in battle. Deklan was trained to fight like William Wallace.

Graduation from SQT meant Deklan Novak was days away from receiving his Navy SEAL Trident. The Special Warfare Insignia, also known as the "SEAL Trident" or in the community known as "The Budweiser" (because the Trident resembled the Anheuser-Busch company logo), recognized members of the United States Navy who have completed the Basic Underwater Demolition/SEAL (BUD/S) training, completed SEAL Qualification Training (SQT) and have been designated as U. S. Navy SEALs. Deklan's Trident ceremony took place in 2008 on the grinder deck at BUD/S, where so many SEAL candidates rang out, quit, and went back to the fleet. The pull-up bars and dip bars lined the concrete-asphalt area as if they had pulled up chairs to lionize the ones who made it and to say…bravo, job well done!

Deklan received six tickets to the Trident ceremony and corralled seven more tickets from graduates that did not have family attending the Trident ceremony. Deklan's family and friends in town far exceeded the number of tickets to the ceremony that Deklan had been able to corral. Deklan's entourage with tickets included his wife, his younger brother DJ, his sister Dana, his mother, his grandfather on his dad's side, Goran Novak, uncles Norman Novak, Roger Novak, Bret Novak, and John Godek (Julia's uncle, former Navy man and American Legion Commander at Deklan's local American Legion Hall, Anthony Godek (Deklan's military mentor, former Annapolis Chaplain, and mom Julia's brother), high school best friends Nathan Garfield, Pete Brunson, and Jack Davidson. Daniel Novak Sr did not have a ticket to the Trident ceremony. Deklan told his father that he was welcome to attend the after-party at the hotel, but there had been limited tickets distributed to the graduates and the tickets went to those people who stood behind him for years. Deklan never hesitated when his father picked up the phone following his arrival in Coronado.

"Dad." Deklan began the soliloquy that he had recited in his head many times. "You were not there when DJ, Dana, mom, and I needed you to be there. You are welcome to come to the after-party at the hotel, in fact I want you to be there, but there were limited tickets available to the graduates. When I had to choose who got the tickets, you drew the short straw fifteen years ago."

CHAPTER FIFTEEN

"Deklan. I am your father. I gave you what it took to get through everything you have gotten through." Daniel replied in defense over the phone.

"Mom got us through the bad days, not you. You had nothing to do with me and my aspirations. Don't give yourself any credit in California or anywhere down the road for that matter. The sweat, blood, and tenacity I needed and found came from my mother." Deklan's anger grew as he departed from his rehearsed text of what to tell his father about not being invited to the Trident ceremony. "You left me when I was eight years old. I am assuming that you have not forgotten that. We saw you whenever it was convenient for you to see us. Do you remember what you told me when I joined the Navy to become a Navy SEAL?"

"Vaguely." Daniel Sr stumbled as he recalled all too well what he told his oldest son when he joined the Navy. "The odds of making it through that program were slim to none, Deklan. I wanted you to go to college." Daniel pleaded his case.

"You told me that I didn't have a chance in hell of making it through Navy SEAL training. I think your exact words were...*why do you want to waste at least a year of your life killing yourself only to find out it wasn't for you.*"

"Your mother didn't think you would get through BUD/S either. Neither did your grandfather, uncles, and cousins. We were worried about you, as parents, that's all." Daniel Sr was grasping at thin air. He decided to lie and change the subject. "Did I tell you, Deklan that I've been sober for seven weeks?"

"Talk to me when you've been a sober man for a year. I'm glad you stopped drinking." Deklan didn't care, knew his dad was lying about his sobriety, and wanted to tell his father that he was over a decade too late for a full reconciliation. "Mom gets a pass. She didn't want me to go to war. Mom didn't want me to get killed. You were drunk when I told you about joining the Navy. You laughed at the idea." Deklan shook his head recalling the exchange. "You fucking laughed when I told you I joined the Navy to become a Navy SEAL. Do you want to attend the party for me or not? I want you to be there, but I'm not begging you to come." Deklan felt the anger fade.

"I want to come." Daniel said defeated.

"Get the details from mom. She knows that I was asking you, but do not grill her or bully her into finding another ticket to the Trident ceremony." Deklan added.

"I won't." Daniel had no choice. The call ended with Daniel Sr agreeing to behave. That proved to be a tall order.

When the Trident ceremony finished on the grinder deck, Deklan spent some time with his SEAL teammates. The beer flowed fast and a happy group of aggressive, amped up, tightly wound new warriors talked about the hell they had been through for the past year. At some point in time a few years back, the traditional neanderthal practice of literally pounding the SEAL Trident into the flesh of the chest at the Trident ceremony was abandoned. Deklan and Deklan's fellow graduates decided ahead of time, to do it anyway. They made an agreement with one of their favorite SEAL BUD/S instructors to do the honors outside of the peering eyes of the brass. The group met outside of the berm, an hour after the formal ceremony. Deklan and his buddies wore their Navy Service Dress White uniforms and had their Trident three prongs unfrogged. Everyone lined up and the BUD/S instructor gave a short speech and proceeded to walk down the line and pound the freshly minted Tridents into each recipient's left pectoral. After the pinning ritual had been completed, the group rushed to the surf zone together and entered the ocean for the first time as Navy SEALs.

Deklan's family and friends went back to the Coronado Island Marriott Resort and Spa. Daniel Novak Sr was waiting at the resort in the restaurant where Deklan's party was booked to be held. Many of the Novak family and friends without tickets to the Trident ceremony had arrived at the hotel, as well. Daniel Sr sat at the bar and was more than sufficiently lubricated when the family arrived from the Trident ceremony. Uncle Roger arrived first and approached his younger brother. The rest of the ticket entourage members arrived shortly thereafter. Deklan was scheduled to arrive at the Marriott after his grinder party ended. Most of the SEALs had family in town, so the grinder party wasn't destined to be an all-night affair. The bartenders at Albaca, the hotel's top dining facility, were suddenly very busy with the Novak family and Deklan's friends.

Daniel Sr had arrived at the Albaca bar mid-afternoon when the Trident ceremony at the base began. Daniel Novak toasted his son and the major-league achievement that he had accomplished. The whiskey told Deklan's father that he had failed as a parent (not a protected national security secret), but that he had tried over the years to make up for the inevitable separation anxiety that was hatched by a divorce. The whiskey clarified the marital split by pointing out how Julia had sabotaged his efforts to be a good father and built roadblocks across every pathway Daniel had attempted to navigate since the divorce. Daniel's irritation mushroomed and he ordered another drink.

"What's up little bro?" Roger Novak slapped his younger brother hard on the back. "Holy shit, Deklan is a certified bad ass. You better be nice to him, bro. You have a Navy SEAL for a son. Holy shit, I can't wait until Deklan gets home on leave. I'll take him with me to all the bars in Chicago and talk shit to everyone."

"Fuck off Roger." Daniel didn't look up. "Deklan wouldn't go with you to the hardware store, much less go out with you for an evening drinking in the city."

"You're probably right, Danny boy." Roger replied with condescending sarcasm. "But at least, I got a ticket to the Trident ceremony. Uh...what happened to your ticket asshole?" The other members of the entourage had arrived by the time Daniel Sr threw the first punch.

Roger Novak was a successful businessman in Chicago. The two brothers did not socialize at all back in Chicago. Roger was married with three high school aged kids and lived in Northbrook, Illinois. Daniel was around periodically to see his kids in Wilmette, but the two brothers remained estranged. Roger never liked the way Daniel walked out on Julia, despite his elation that Julia and the kids were away from the alcoholic disaster, more commonly known as Roger's younger brother. The first punch was on target and landed with the power of a stunted Ken Norton early round assault. Roger went down hard, and blood ran from his mouth like the Arkansas River rushing past Buena Vista headed east towards Pueblo.

"Mother fucker." Roger yelled from the floor of the bar. Daniel was standing like a triumphant gladiator, eyeing his fallen prey. Surprise punches with some juice always did major damage. Bret Novak and Norman Godek raced in to join the fray. The remaining Novak and Godek family members and hotel guests stood stunned as they gasped and watched what suddenly became a scene from the set of Animal House with a guest appearance by the ghost of Tyler Durden. The battle odds instantly became three against one. Daniel Sr realized his backup plan was nonexistent.

Roger got up like he was on fire and went after Daniel by tackling his brother. Customers at the Albaca bar scrambled like a school fire alarm had been activated. The two falling adults crashed into a high-top bar table with an older couple admiring the view of the San Diego skyline. Suddenly, there was a physical altercation playing out inside an establishment ill-equipped to handle a fight of any kind. Dozens of shocked patrons hurried to locate a pathway to escape the violent melee. The unmistakable sound of an adult fist crashing into a face caused panic in the eyes of the happy hour guests. The Marriott and the Albaca bar were a stone's throw away from the site of the toughest training in the world that produced the toughest men in the world, but the nearest bouncer was down at McP's Irish Pub on Orange Avenue.

McP's Irish Pub was the most well-known watering hole in Coronado. SEALs often chose one of two other bars on Orange Ave for their beverages and their caveman competitions. Danny's was considered the most real SEAL bar on the island and the place to go when there was a SEAL celebration. One could always find a fireteam of SEALs at the end of the bar inside Danny's. Little Bar was a hole-in-the-wall with a couple pool tables, dart boards, a deep jukebox, and a corner to slip away and be invisible. Both bars were more popular with Navy SEALs than McP's. McP's Irish Pub was too often packed with dozens of chicken necking tourists and sun worshippers looking to catch a glimpse of a tatted-up meathead with a thousand-yard

CHAPTER FIFTEEN

stare. The chaos from the Silver Strand Training Complex South and the associated testosterone confrontations generally landed outside of the luxury hotels serving Coronado. The bartenders at Albaca called for backup to the front desk and sheepishly attempted to intervene.

Bret Novak, Daniel's younger brother, and Norman Godek, Julia's younger brother did not hesitate to jump into the fray. Everyone involved was feeling no pain. The graduation party on the grinder was hardly a gathering to celebrate Teetotalism and the advent of abstinence. Suddenly, Daniel Novak was in a brawl he started, but the odds of a victorious retribution dance dwindled down to nonexistent. Julia's brother, Norman, was unemployed periodically and generally lived from paycheck to paycheck. Norman took blockhead jobs in the city from working as a Security Guard at Macy's on Michigan Avenue in Chicago to working as a bouncer for Park West, a premier concert venue located at Armitage and Lincoln in Chicago, only to get fired for showing up drunk. Getting the opportunity to kick Daniel's ass was a gift from above for Norman. Bret Novak had been a rugby player in college, remained in great physical condition, could have posed for Shred Magazine, and dabbled in every extreme sport on the planet. Bret Novak was a force to be avoided.

The fight escalated quickly. Daniel threw wild punches at three men, while they took turns rearranging the flesh attached to Daniel's face. Brett Novak couldn't stand his older brother and rocked Daniel multiple times like he was Mike Tyson pummeling Trevor Berbick. Brett held his hands high in a peek-a-boo style mimicking the former heavyweight champion. Blood covered the floor at the Albaca bar. Tables had been knocked over. Glasses and food plates were strewn around like there had been an earthquake. Guests were screaming as they ran for the hotel lobby. Fronts desk clerks, the Concierge and three bell boys ran into the Albaca space to help quell the turmoil. Daniel Sr stood his ground admirably, but got his ass kicked by family members that seemed to enjoy every second of the altercation.

In the end, Daniel Sr landed in the emergency room at Sharp Coronado Hospital with the Coronado Police Department waiting to arrest the newly admitted patient as soon as Daniel's head was stitched up. The Novak and Godek families were asked to leave the Coronado Island Marriott Resort and Spa immediately, regardless of who started the fight. Goran Novak booked the entire family entourage into the Hotel del Coronado on his dime and the festivities barely skipped a beat. New memories were added to the bar banter archives for the Novak clan. When Deklan arrived at the Marriott, he was told of the Albaca brawl and left to meet his displaced family at the gold standard for hotels in the state of California. Deklan didn't process what had happened until he saw his father's face later that evening. Obviously, Deklan knew from the moment he heard the seven weeks sober line, that his father was lying. Daniel's face was slam-dunk evidence.

Grandpa Goran Novak paid to stop the Marriott management from pressing charges. The police spoke to Daniel at the hospital, gave him some sound suggestions pertaining to his future behavior on the island and the CPD left the hospital without making an arrest. Daniel Sr left the hospital with a face that resembled Tommy Hearns after three rounds with Marvin Hagler on April 15, 1985. The Hagler/Hearns fight had been dubbed "the most electrifying eight minutes in boxing history." Daniel Novak wore a blackish/purple hue on both cheeks, had bandages on both hands, took twelve stiches to his mouth and his lips looked like they had received an overdose of Botox. There was a gash over Daniel's left eye and his right eye was swollen and appeared closed. The Coronado Police Department officers called to respond to the fight at the hotel and who subsequently accompanied Daniel to the hospital, stood and watched their newly freed suspect signal a cab waiting in the hospital's parking lot. There was a slight shit-eating grin on the face of Daniel Novak Sr as he slowly got into a taxi.

At 77 years old, Goran was the family patriarch. Goran was a real estate entrepreneur in the fifties and sixties in Chicago. Goran bought property inside many destressed areas within Chicago. Goran bought buildings in the Near West Side communities near the old Chicago Stadium. Goran Novak was advised to purchase property elsewhere by a plethora of lawyers and investment bankers at the time. Goran

CHAPTER FIFTEEN

ignored the advice and bet on Michael Jordan and the Chicago Bulls to pull the West Side out of poverty. In the mid-late nineties, Chicago's West Side began a shift into the new hub for wealthy white residents created by men like Goran Novak, fueled by forces like rapid gentrification, corporate investment, and the unequal distribution of city resources. Goran Novak became a very wealthy man because he bet against the odds and the advice of his well-paid team of financial advisors.

Deklan Novak enjoyed the family reunion on his behalf. Deklan's father had already begun to spin the story of the Marriott brawl while sitting at the Babcock & Story Bar inside the Hotel del Coronado. Deklan felt a deep sense of pride as he stood among the dysfunctional components of his lineage. Most of all, Deklan enjoyed reconnecting with his younger brother, DJ.

CHAPTER
SIXTEEN

2008...DANIEL NOVAK JR (DJ)

DJ WAS sober and had been for some time when Deklan mailed out the official Trident ceremony invitations to his family and friends. Novak had 32 friends and family arriving in Coronado the day prior to the ceremony. DJ was sober if one considered a steady dosing of methadone sober. But DJ was closer to being himself than he had been in a while. He struggled hard off and on with drugs and alcohol since high school. When sober for a period, DJ was a brilliant, beautiful human that Deklan needed to spend time with. Deklan was so thankful that DJ had accumulated some time clean and seemed very stable in his sobriety for the first time in years. The brothers had many extended, personal conversations during the months prior to the graduation ceremony. Deklan spun tales to DJ about Southern California and the insane journey he was on. Daniel Jr was fascinated by the training tales and his older brother was so hopeful for the next life chapter to be written for DJ.

Perhaps as much as the graduation ceremony itself, Deklan looked forward to receiving DJ at the airport and showing him every spectacle in the Crown City, and the secular society that was thriving in and around the San Diego Bay. Deklan could tell immediately that DJ wasn't exactly clean. Daniel Jr's methadone left him with a slightly glassed over and hazy look in his eyes and he would occasionally lose track of their conversation. But that was a far cry from the days of DJ nodding out mid-sentence at the cash register of an Amoco station.Deklan didn't care, DJ was trying, and he was at least close to his real self. Good enough was often sufficiently better than expected.

The first thing DJ and Deklan did was grab a beer in the gaslight district of downtown San Diego, then they crossed the historic San Diego bridge onto Coronado Island. Deklan enjoyed from the driver's side of the front bench seat of his 1995 Ford F 150, the amazement and excitement in DJ's eyes. It reminded Deklan of the first time he saw and drove over the same bridge a year and a half before, when a young sailor arrived for BUD/S training. The pair drove down Orange Avenue, then to the beach front, then Deklan took DJ to his favorite place, The Little Club to celebrate. Deklan wasn't a SEAL yet so Danny's and McP's were pretty much off limits for different reasons. They drank beer after beer and played a half dozen games of pool together. DJ was an exceptional pool player. DJ beat Deklan in five of six games, but that did not matter because Deklan was so happy to see his brother, and so happy to see the way he strutted about bursting with pride that his big brother was about to become a Navy SEAL. The confidence and joy in DJ's eyes had been absent for such a long time. Deklan, like so many relatives of addicted family members, mistakenly held that consuming a few beers was harmless.

The brothers capped the evening off at about 10 pm and headed back to Deklan's house. DJ stayed with Deklan that first night, as the rest of the family and friends arrived the next day, DJ would stay with them at the swanky Marriott Resort with its spectacular view of downtown San Diego across the bay. Deklan showed DJ to his room and gave him a bear hug, lifting him off the ground while looking into his smiling face, Deklan told DJ how proud he was of him and how much he loved him. DJ closed the door behind Deklan and hit the rack. The next day was the last day that Deklan Novak would be a Navy SEAL candidate, and he had 32 family members and friends coming to Coronado. It was going to be a long day.

CHAPTER SIXTEEN

Many years before:

"What are you some kind of faggot!?"

Daniel Novak Sr liked to call Deklan Hotshot or Player, but he really liked tearing into Daniel Jr. Deklan's father tore into his younger brother so often by inquiring if DJ was a little fairy boy or a faggot. DJ was a sensitive young kid. He liked to play and pretend and dream. Deklan was a heat-seeking missile, wired to jump off any roof in the neighborhood, steamroll the neighborhood kids in a flag-football game, and spend hours watching the WWF on television and specifically, Mark William Calaway (aka The Undertaker). Daniel Novak Jr was softer as a child than his older brother, kind and sweet with no ill will or ostentatious tendency towards violence. Daniel Sr didn't like that. He didn't like that innocence. Daniel Sr. didn't like anything that could have been construed as feminine or emotionally based. Hugs, physical affection, holding hands, affirmations, and kindness, were all father to son outward expressions of affection that Daniel Sr never practiced. Daniel Sr didn't like or approve of his young son, Daniel Jr. Daniel Jr or DJ (the nickname given to her baby boy by Julia) was great at sports, but he didn't care about them, did not care about competition, and was not obsessed with winning and losing. Daniel Jr was good at almost everything he attempted, but where he excelled the most was as an artist. DJ had the natural talent to take a piece of paper, grab a pencil and turn the paper into a stunning sketchable reflection of the imagination within his mind's eye.

Deklan's younger brother was a creative genius and that angered his father. Daniel Sr regurgitated a mantra daily when assessing his younger son:

"What are you some kind of little faggot sissy boy? Why don't you play harder out there like your brother?" Daniel Sr would say cornering the 10-year-old little boy in the car on the way home from football practice. Coaches often queried DJ's father about why DJ wasn't more like Deklan. The coaches were not slamming DJ. They were complimenting Deklan, but the insensitive inquiries acted like jet fuel on a Butane lighter for DJ's delusional father.

Daniel Jr's father loved to call DJ a faggot, as if he believed the fear of being seen as gay or a sissy by his dad would drive Daniel Jr to accomplish great masculine feats of strength. Daniel Novak Sr must have believed on some level that degrading a child whose passions included some less than perceptual macho passions would eventually alleviate all of life's pain and suffering.

Daniel Jr was bullied by his father but ignored the barrages mostly because Daniel Sr lived outside the family home. DJ was kind, thoughtful and sensitive. Daniel Jr was loved by everyone, and he was a creative tech genius. DJ learned how to program computers on his own mastering computer language C, C++, and Java from reading books and writing his own programs at the age of 21. DJ was fluent in the guitar, piano, harmonica, and played the best guitar hero Deklan had yet to see. Daniel Jr could paint, draw, write poetry and compose songs complete with music and lyrics. Deklan's younger brother was good looking and charismatic. DJ always caught the eye of the best-looking girl in the class no matter which social club she came from, and he had no problem talking to any of them.

Although DJ never considered himself an athlete, he was instinctually good at anything physical. DJ was a brilliant boxer, with an uncanny ring savvy, yet he was never formally trained in the art of pugilism. When Deklan and DJ sparred, DJ often struck his older brother two or three times before Deklan could effectively counter or back out. The untrained timing was impeccable. The suffocating pain that he carried was too much even for all those beautiful talents to overcome. Interpersonal context was the social multiplex that defined most individuals and DJ's delicate sense of self was fractured.

The scales began tipping early in high school for Daniel Jr when he was introduced to weed. Slippery slopes eventually spilled all the way over into the darkness. By 23 years old, Daniel Jr was a journeyman heroin addict, possessed an addictive personality, experienced blackouts, and was an alcoholic. If it wasn't for coming from a family of alcoholics and drug addicts, it would have been unfathomable how such a gifted and talented young man could be choosing to throw his life away the way that he was. The questions of how he could be sabotaging his life were rarely asked by the man who wanted little to do with someone so weak. The sad revealing truth, as expressed by Daniel Sr, on the way to take Daniel Jr to another addiction treatment program told it all:

DJ's father said, "We are all (the family) lucky that our drug of choice is alcohol. Heroin is a death sentence." Deklan wanted to break his father's jaw, but there was a time and a place for everything. The present day was assigned to help DJ. Jaw shattering timelines were to be addressed down the road. "There's no longevity." The idiocy from DJ's father continued. "We can drink for the rest of our lives. But DJ is either going to end up in jail or dead or both."

Julia Novak recalled Daniel Jr telling her six months before he died, "You know the only time that I feel ok about myself is when I am high on heroin."

DJ never could see how amazing he was or how talented he was. Addiction destroyed a brilliant and beautiful young person. Julia fought like hell to rescue her son and Daniel Sr found solace from his own inebriated insecurity by pointing out how many poor choices DJ had made in the past and continued to make during his emotional war. DJ never felt brilliant and beautiful. Addiction was loneliness, regret, defeat, and hopelessness, that arrived like a calculated boxer attacking his opponent with crisp jabs and bone-crushing haymakers. The contrast between the never-ceasing self-loathing and the numbing high of opioids was the closest thing DJ could feel to love, peace, and happiness. Unfortunately, that kind of peace came at a great price. The damage became apparent gradually and there was nobody in the family equipped to heal those types of wounds. DJ felt alone and he was alone.

In retrospect, Deklan knew that no one helped his little brother, including himself. Daniel Jr's father used "tough love" to isolate the issue. Daniel Sr kept hammering home the message to his struggling son: "I never taught you to behave that way." The irony was palpable.

"Tough love" used by a man who ran behind a bottle of Jack Daniels or Jim Beam for his entire adult life and sought out the wisdom provided by a Kentucky bourbon for his dalliances in the psychodynamic perspectives towards understanding addiction. Deklan wanted to help DJ, but he didn't know how to help. With Freudian uncertainty and stumbling good intentions, Deklan meticulously kept his personal triumphs away from the family as much as possible. The MMA success, joining the Navy, and BUD/S were seemingly the arduous actions of a brother to be admired and looked up to for a well-adjusted, wide-eyed younger brother. The MMA success, joining the Navy, the improbable graduation from BUD/S, and the Navy Seal Trident graduation ceremony were toxic outliers to the mental stability of a young man searching for his own grinder deck to stand on.

The periodic visits to sobriety lasted no more than a few months each time. The methadone treatment alternative proved to be equally unsuccessful. Methadone eased the tension but didn't have the punch needed to shield DJ permanently from the always looming cycle of fear and self-loathing. For the addict, the terrors always lurked behind the mask of shared family triumphs, maliciously probing the cloak of self-esteem. For most hailed accomplishments in the traditional family dynamic model, shared family graduations, engagements, promotions, or retirements generated the efficacy on Sunday afternoon to overindulge at a party celebration and battle the guilt during the next day from too many calories and too many cocktails. Triggered by deep unhealed trauma, Daniel Jr found the shared family triumphs or watching his many friends succeed in life was the catalytic converter to jumpstarting the downward cycle back to the needle. DJ searched for the answers to questions he couldn't understand while a crushing wet blanket of fear smothered any semblance of self-esteem and pride. DJ's father chose to bypass the examination. DJ's older brother had been away much of the time and as scared as Deklan was for his brother, Deklan hoped his impossible dream of becoming a Navy SEAL was subconsciously the right medicine that his brother could use as inspirational fuel, charting a flight plan to say that anything was possible. The medicine never got delivered and Deklan knew he had failed in the same manner as their father. Daniel Jr battled alone and lost.

Three days prior to DJ's death, the brothers spoke for the final time. Deklan had been assigned to SEAL Team One. Daniel Jr absolutely adored his older brother and all of Deklan's accomplishments. DJ wanted to be like Deklan but saw no path. They talked several times a week. DJ was a few months sober and that made the relationship closer and stronger. Deklan began to imagine a normal life for his brother, and he felt DJ could handle a bit of responsibility. Deklan lived in Coronado, CA and DJ lived in Wilmette with Julia. Deklan asked DJ to tie up a few loose ends back home and gave his younger brother a handful of tasks to keep him busy and help Deklan recover some business assets he had left behind with his former partner. Deklan's little brother accepted enthusiastically. Deklan detailed the tasks in simple terms and explained the contacts involved. A couple days later, they talked again. When DJ was asked about the response from the two contacts he was supposed to reach out to, DJ admitted that he hadn't done the favors that Deklan had asked him to complete. Deklan had not had a good day with SEAL Team One. The new guy hazing and bullshit SEAL Team initiation, had been especially thick on the day Deklan and DJ spoke. Deklan's blood had been boiling most of the day. Arrogantly and foolheartedly Deklan thought DJ was looking for a way to blow off his Navy SEAL brother. I'm trying to help him God dammit, Deklan thought. Before hanging up the phone, the last words that Deklan said to DJ were the last words he ever spoke to his brother.

"You lazy fucking piece of shit, you can't even do the easiest fucking thing I asked you to do."

Three days later Deklan received the shocking life altering call. The phone buzzed in his pocket, and Deklan knew exactly what he was about to hear when he saw the incoming call was from his sister, Dana. Dana's barely coherent hysterical news needed no translation or clarification. Deklan knew that Daniel Jr was dead. DJ was found in the basement of their mother's home with a needle half plunged and still hanging from his lifeless arm. DJ had died alone and had been dead for several hours before his overworked mother found him after a long day. There was nothing Deklan could do to console Dana. They both hung up in tears. Julia and Dana watched the ambulance leave the house with Julia's baby boy and Dana's big brother gone forever. Deklan was on the 17th hole of Coronado's Naval Air Station North Island Sea N' Air golf course with JD Christie, Deklan's buddy and the owner of Coronado Sports and Spirits on Coronado's Orange Avenue. SEAL Team One's new breacher and Team gopher, Deklan Novak never completed the golf round.

CHAPTER SEVENTEEN

2014...BEHR'S UNANNOUNCED VISIT TO STEAMBOAT SPRINGS

HOTEL STEAMBOAT SPRINGS was located a five-minute walk from the Steamboat Ski Area. The three lower mountain lifts serviced most of the green and blue runs. Blue-black runs were accessible from the Pony Express lift on Pioneer Ridge. Behr would not be testing his mogul skills during his stay in Steamboat Springs. Behr Thomas used travel miles to offset the January high cost of a Steamboat Springs hotel.

Mount Werner is a 10,570-foot peak located in Routt National Forest, 4.6 miles east of downtown Steamboat Springs, Colorado. Formerly known as Storm Mountain, it was renamed in 1965 in honor of Buddy Werner, an Olympian from Steamboat Springs who was killed in an avalanche in Switzerland in 1964. The Steamboat Ski Resort opened in 1963 and operated on 2,965 acres of the mountain. The resort averaged 334 inches of snow per year over the past ten years, some of the highest snow totals in Colorado. The mountain was visible from the window of Room 440 at the Hotel Steamboat Springs on Mt. Werner Rd. Behr Thomas looked at the snow-covered peak in January from his hotel window.

The hotel had been a mainstay in Steamboat Springs for decades.

Behr pictured the town when his parents wore bell-bottoms and hitch-hiked down Interstate Hwy 70 to Colorado from the University of Kansas for ski vacations during their college days. The nostalgia was not entirely born from warm and fuzzy memories. The Hotel Steamboat Springs smelled like an ash tray. Nicotine hung inside the three-story lobby with arching pine log beams like the humidity in Maldives. The never-ending carcinogenic residue painted images of young models smoking amid the pages of Life Magazine fifty years ago. The malodorous memories from Philip Morris International had taken permanent residence inside the Hotel Steamboat Springs. Hilton and Marriott had recently built five-star hotels next to Mt. Werner. The Hotel Steamboat Springs hung on to their dated lineage like a Catholic Irishman held onto his Gaelic origin. The music in the hotel's lobby when Behr checked in was *Locomotive Breath* from Jethro Tull's 1971 Aqualung album.

From the rustic chandeliers to the dated Italian restaurant and through the Saddle & Cinch coffee shop, where the discolored ceiling tiles and ventilation ducts looked like they hadn't been touched since Nixon resigned, the 70's were revived. The hotel's prices per night reflected the accommodation. The past presence of cigarettes was unmistakable. Although the structure had undergone numerous remodeling phases and renovations, there was only so much lipstick that could be smeared on a pig before one realized the beast was still a pig. New paint and newer televisions every ten years was about the extent of most Hotel Steamboat Springs remodeling projects. Behr's hypersomnolence brought a burning sensation to his nose, a dry cough, and watery, itchy eyes every time he entered the building. Dry cleaning, steam cleaning, commercial washing machines, and exorcisms were useless in removing nicotine from hotel curtains and hallway carpeting. Nothing masked the smell of cigarettes from fifty-plus years of respiratory exhilaration among the original white powder junkies wearing Descente fitted ski jackets and lugging their K-2 skis.

Behr Thomas came to Steamboat Springs to see his daughter Kaley after the 2013 college football season had ended. The Wildcats from

CHAPTER SEVENTEEN

Northwestern finished 5-7 in the Big Ten Legends Division. It was Lance Adam's and Behr's eighth season with Northwestern. The holidays had been frustrating for Behr. Kaley was scheduled to visit during the week after Christmas from 12/27/2013 until 1/2/2014. Kaley hadn't seen her father in almost a year. Behr had tried to abide by the court's custody agreement, but when Raya kept using Kaley's undetermined illness as an excuse to prevent Behr from seeing his daughter, the red flags began to rise. Behr Thomas was not a child psychologist or social worker, but he knew when something wasn't right. If Raya and her dirt-bag, ex-con boyfriend were abusing Kaley, it didn't show beyond her weight and that may have been due to her health issues over the past year. Kaley did not have a mark on her and appeared untouched. Thank God, Behr thought for his daughter and her unrelated live-in guardian. Behr often imagined someone hurting his little girl and what he might be capable of doing to such an individual. Fathers almost universally traveled that road at one time or another.

The drive scenery to the mobile home park quickly changed from resort town to mobile homes and abandoned cars. The address was depressing to look at because Behr knew that his daughter called the place her home. Behr knocked on the front door of the trailer. There was a small bench on a smaller porch next to the front door of the single-wide mobile home. Behr called the 8th Street Steakhouse and asked if Raya was working. He told whoever answered the phone that he was an old friend and was going to come by and surprise Raya, so he asked the person not to say anything. Raya was at work. Kaley had to be home. Raya had told Behr that she was home-schooling Kaley until they could afford to send Kaley to Emerald Mountain School, a private school in Steamboat Springs with curriculum from K-12 grades. Tuition at Emerald Mountain school started at $15,300 annually. Behr was hoping Carmen Gallardo was also home. Behr got his wish. Gallardo opened the front door, and the two men stood there silent for too long. Neither man wanted to talk first. Sergio Leone inspired silence eventually ended.

"Raya is not home." Gallardo finally spoke. "She's at work. She works at the 8th Street Steakhouse."

"I know where Raya works, and I know she isn't home." Behr answered back.

"Then why are you here?" Gallardo asked and the tone was obvious.

"My daughter lives here." Behr said.

"You don't."

"I'd like to see my daughter." Behr demanded.

"No." Gallardo barked. "The custody agreement between you and Raya does not allow for unannounced visits. This is an unannounced visit, asshole. Take off and make the arrangements in advance when you want to see Kaley."

Gallardo was a decent sized guy, but no match for Behr Thomas. Thomas knew that and so did Gallardo. Gallardo was baiting Behr Thomas.

"Is Kaley here?" Behr asked.

"Like I said. Take off my friend and make arrangement in advance when you want to see your daughter."

"Be careful, Mr. Gallardo." Behr spoke calmly as he started to leave. "When a bear is hibernating, leave it alone. I hope you considered the end game to what your tongue started."

"Are you threatening me?" Gallardo asked.

"I am enlightening you." Behr replied.

"Like I said twice, plan to see Kaley in advance with your ex-wife. Now, fuck-off." Gallardo closed the door and didn't wait for a response.

Raya Nolan Thomas was not happy to see her ex-husband when Behr wandered into the 8th Street Steakhouse. Behr sat at the bar and ordered a beer. Raya took her time to walk over and find out why Behr was in town and what he wanted.

"I would say this is an unexpected pleasure if it was." Raya smiled sarcastically when she finally addressed her ex-husband at the bar. "What do you want, Behr?"

"You've lost weight." Behr commented on his ex-wife who

CHAPTER SEVENTEEN

normally looked great. "I had the pleasure of meeting your boyfriend a few minutes ago. Very personable. I can see why you are attracted to him."

"You are not supposed to show up unannounced." Raya reminded Behr. "What did you expect?"

"I was in town to visit a recruit. Came up as a last-minute thing." Behr lied.

"Bullshit." Raya shot back. "What's the kid's name and what high school did he go to?"

"The kid's name is Jason Stillwater. He plays defensive end and is graduating from Steamboat Springs High school in June. He's 6'3" tall, weighs 295 lbs. and runs like an Olympic sprinter. He ranked 24th in the ESPN Top 300 recruiting database for this year. Jason Stillwater was in the top 20 for Max Prep recruiting rankings and in the top 30 for USA Today high school football recruit rankings. He's a legit player. Bama wants him. Clemson, Ohio State and Michigan want him, but his parents went to Northwestern, and he wants to come to Evanston if everything fits his plan. I had to come here, and I took a chance to see Kaley. That did not happen." Behr had prepared an excuse to show up. Divorce elevated dishonesty to a second nature phase.

"You cannot show up unannounced, Behr." Raya explained again.

"I am aware." Behr responded. "When can I see my daughter? I am not here very long. Since I am here, I am not leaving until I see Kaley." Behr was on shaky ground. A judge would not like to hear about a threat, regardless of how it was presented. A threat was a threat. Custodial awards from a divorce settlement were not the lines in the sand to cross.

Behr waited in the restaurant at his hotel for Kaley and Raya to show up. Raya had agreed to bring Kaley to see Behr early on the next day, when Behr was scheduled to fly back to Chicago. Behr's flight was later in the day. Raya's insistence on Behr not returning to their trailer was suspect to Behr but a physical confrontation with Gallardo was not going to help anyone, especially Kaley. Behr, in the back of his head, wanted to return to the trailer to see his daughter. He wanted to stand face to face and toe to toe with Gallardo again. Behr wanted to hurt Gallardo in the worst way. Raya agreed to bring Kaley to the Hotel Steamboat Springs. Behr settled for what he could get without a confrontation.

Behr waited in the coffee shop at the resort hotel called Saddle & Cinch Café and Bar. Behr drank black coffee next to a massive fireplace while looking out at the base of Mt Werner and the crowded gondola lift lines. The base of the mountain was packed with people on a cold January morning. Skiers and snowboarders waited to board the passing glass enclosures for Steamboat Gondola and Christie Peak Express Gondola.

Raya and Kaley walked into the restaurant holding hands. Kaley seemed distant, as if she was attached to her mother but was staring at the ground. Raya looked mad, as usual. Raya and Kaley joined Behr at his table. Kaley was bundled up from the harsh Northwestern Colorado wind in January.

"Hi, honey." Behr leaned over to kiss his daughter, but she kept her head down and appeared oblivious to the attention. "Are you hungry, Kaley? The food here is great. Pancakes are the best anywhere."

"She'll have some cereal, Behr." Raya chimed in before Kaley could respond. "You do remember that Kaley has been having some big issues with her digestive system." Raya sat back in the booth and did not remove her scarf or her jacket. Kaley kept her jacket zipped up as well.

"That's good." Behr did not want to argue or battle Raya in front of Kaley. Raya had a year since their divorce to fill Kaley's head with anti-daddy propaganda and any show of frustration or anger by Behr would only fuel the lies. Raya ordered cereal for Kaley and coffee for herself.

"You're not eating?" Behr asked Raya.

"Nothing gets by you." Raya snapped sarcastically.

"Sweetie, have you made any new friends this year here in Colorado?" Behr turned and addressed his daughter. Kaley did not respond immediately.

"She's been pretty sick this year and we have kept her at home until we can know for sure what is wrong." Raya chimed in. "We cannot go around infecting other children, now, can we? That isn't fair to the other kids in school or in the neighborhood."

"That makes sense Kaley." Behr reassured his daughter although he wanted to slam his ex-wife's head into the table. "Have you had a chance to go skiing or sledding here? I can't imagine how many places there are to go ice skating, or skiing, or sledding. This is truly a winter wonderland."

"I just got done telling you that Kaley has been sick this year. Why ask her if she has been skiing or sledding when you know that she has been sick so much this year?" Raya asked as confrontational as one can get without spitting.

"Do you allow Kaley to speak, Raya? Or is she too sick to talk to me?" Behr had enough.

"Did you ask us here to start a fight?" Raya reached for Kaley's hat. Behr knew this was over.

"Let's review what I did not ask you for. I did not ask you to come here. I wanted to see Kaley by myself, but you prevented that from happening. You prevented Kaley from coming to Chicago over Christmas. When I finally get to see my daughter, you do not allow her to speak. What are you afraid of, Raya?" Behr knew they were about to leave.

Behr had researched child abuse before he traveled to Steamboat Springs. Kaley's aborted visit to see her father over Christmas convinced Behr that someone had a reason to stop Kaley from coming to visit her father. Raya had agreed to the visitation arrangements that the non-custodial parent (Behr) was entitled to. Behr had not fought the sole physical custody determination made for Raya. Their marriage was over, but they had a child together and they had agreed that Kaley would always love her daddy. Therefore, Gallardo was the problem. Behr had determined that Kaley was in imminent danger of being harmed by Carmen Gallardo. Raya brought their daughter to Colorado and placed her in a household with an ex-convict capable of anything. Gallardo already had been convicted and jailed as a drug dealer and for assault.

An attorney friend from Northwestern's Law School told Behr that no court would change the custody awarded from a hunch or a father's intuition. "Jeopardy or Imminent Danger" had to be proven, and the quest was often long and very expensive. Behr was told to visit the girl to access her living conditions firsthand and gather any evidence if there was evidence to gather.

The breakfast reunion lasted less than thirty minutes.

"Thank you for breakfast, Behr. Kaley, thank your father for breakfast." Raya announced as she was sliding out of the booth.

"Thank you." Kaley replied. There was no follow-up. There was no mention of "daddy". Kaley had been coached big time. Behr wanted to grab his daughter and bring her back to Chicago with him, but that would have to wait. Kaley and Raya left the restaurant. The interaction during breakfast was nonexistent. Behr had five hours before he was scheduled to leave for the airport. Behr departed from the Steamboat Springs airport in Hayden, Colorado at 6:00 p.m., changed planes in Denver and arrived at O'Hare International Airport in Chicago at 2:00 a.m. There would be one more visit to the Dream Island Mobile Home Park.

CHAPTER EIGHTEEN

2014...DEKLAN NOVAK AND BEHR THOMAS

BEHR THOMAS and Deklan Novak were set to meet for breakfast at Walker Bros., The Original Pancake House on Green Bay Road in Wilmette, Illinois. The restaurant was five minutes from the Northwestern campus and Behr's office at Anderson Hall/Burton Academic Advising Center. Most of the university's athletic offices were in Anderson Hall, just northeast of Ryan Field, where the Northwestern University home football games were played. Walker Bros., The Original Pancake House opened their first restaurant in Wilmette in 1960, serving amazing morning fare and famous around Chicago and the North Shore for the Apple Cinnamon Pancake.

Amber and Deklan were separated by the winter of 2014 for what was originally deemed a trial separation. Amber had taken the kids back to her mother's house in Kenilworth, not far from the restaurant where Behr and Deklan were set to meet. Reconciliation was on the table for Amber and Deklan, but the moving parts seemed to change daily, and Deklan's own demons weren't helping. Captain Morgan was not a beacon of direction at the time. Deklan was still working at Great Lakes Naval Recruit Training Command (RTC Great Lakes) as an active-duty Navy SEAL teaching recruits how to prepare for

BUD/S training, in addition to his obligations as a full-time student at NU and a football player. Big Ten football was a year-round commitment. Deklan remained living at the home that he and Amber bought when the assignment to RTC Great Lakes was finalized following Deklan's final deployment overseas in 2012.

Behr had arrived first and was practicing what he was going to say to Deklan Novak. On the one hand, Deklan Novak was a student and football player under his tutelage. On the other hand, Deklan Novak defined a level of discipline and overcoming obstacles that Behr sought to find. Teachers and coaches at the university level were required to keep their relationships with students purely professional. Most often, the boundaries crossed were sexual boundaries between a teacher or a coach, and a student. Behr and Deklan were clearly not involved physically. Coaches were tutored to not misuse their position of authority by using the power of their position to take advantage of a student. Behr was not taking advantage of Deklan in any way. Behr was reaching out to Deklan for his advice based solely on the extensive life experiences that Deklan possessed. Coaches and teachers were taught not to allow their emotions to influence their behavior around their students. Behr was emotionally defeated and sought out Deklan for reasons he was embarrassed to admit. Behr practiced what he wanted to say before Deklan arrived. The busboy came by every seven seconds to refill Behr's coffee cup. Behr waved his hand over his cup for the fourth time, more annoyed at himself for being weak than he was at the redundant service.

Deklan walked into Walker Bros. and spoke to the hostess, who pointed Deklan in the correct direction. Behr stood up when Deklan approached and wondered if that was overkill. Walker Bros. restaurant was last remodeled in the eighties. The deep wood booths varied with an eclectic composition design, and were all made of expensive cherry, Mahogony, and walnut planks. The ash and maple walls accented the intricate textured Baroque stained-glass booth dividers, and the eight large hand-made Toscano Tiffany-Style stained-glass chandeliers hung like priceless paintings within the dining areas. Brass railings with velvet ropes marked the waiting zones that were packed on every weekend morning. Lines formed early at Walker Bros. and stretched around the building into the parking lots on Saturdays, Sundays, and most holiday mornings. Deklan sat down and the two men shook hands.

"Hey, coach. How was Colorado? Were you able to spend time with Kaley?" Deklan asked politely but vaguely knew why Behr had asked him to meet at Walker Bros. The server brought Deklan some coffee. Black was fine.

"Good to see you, Deklan." Behr replied. "I'm still reeling from the seven-game losing streak we had last season. At least we ended the year with a win over Illinois in Champagne."

"Everything is a learning experience as far as I'm concerned." Deklan observed. "I know this team is headed for great things soon. I could feel it in every player and coach after the season ended."

"I agree with you, but my downward spiral I suppose has to do with my divorce and Kaley more than a seven-game losing streak." Behr got right to the point. "Kaley has been sick since she and Raya moved back to Steamboat Springs. I needed to see Kaley in person, so I used a recruiting trip as an excuse and went to Steamboat Springs unannounced and showed up at the trailer park where Raya and Kaley live with Raya's boyfriend. Carmen Gallardo is Raya's live-in boyfriend. Gallardo is an ex-con, convicted on drug charges and an assault. He did time for both. Great role model for my daughter. I knocked on the trailer front door and Gallardo was home. Raya was not home. The prick would not let me see Kaley. I don't know if she was home at the time I got there. Gallardo barked up something about my showing up unannounced, was a violation of the divorce agreement. He told me to make an appointment with Raya."

"Was he right?" Deklan asked bluntly.

"Yes, he was correct but why prevent a father from seeing his daughter if you've got nothing to hide.?" Behr asked.

"There are many reasons why her boyfriend may resent you. Raya has had over a year since the divorce, right?"

"Right."

"Raya most likely has filled this guy's head with your shortcomings as a husband and as a father. Raya has had more than a year to bad mouth you and justify her actions in leaving you and taking Kaley."

"The guy is bad news all around." Behr commented.

"Can you regain custody if you go to court and tell the judge that Raya moved in with an ex-convict and those circumstances, not known at the time of the divorce, are certainly not in the best interest of the child?" Deklan asked.

"I made a call after I found out Raya was living with this guy." Behr said. "My lawyer told me that I can petition the court to become the custodial parent, but that usually takes 2-3 court appearances and can take up to 18 months to resolve. The court will evaluate the nature of the ex-convict's crimes, the length of time since the convictions, the ex-convict's rehabilitation efforts and if the child is in jeopardy. My lawyer told me that the court may increase my visitation rights first, but since I live 1800 miles away, that would be a moot point. I need physical proof that Gallardo is harming Kaley. I don't have that. Kaley will be lost to me in 18 months." Behr's eyes were fixed on the table.

"No doubt, coach. You need a better live-action plan, a flanking maneuver to attack from behind the enemy's front line." Deklan added using military combat terminology. "Popping in, is not a plan. I'm sure you didn't expect the live-in boyfriend to shake your hand and welcome you into a kumbaya society hug fest." Deklan raised his eyebrows to hammer home the point he was making. "Did you see Kaley at all?"

"Yes." Behr explained. "Raya brought Kaley to my hotel for breakfast on the morning before I flew out. They stayed for less than 30 minutes, and Kaley was super quiet and barely talked to me. Raya seemed to have a spell on Kaley. My ex-wife reminded me that Kaley had been sick and that was why she was acting so reserved, but there was something else, Deklan. I couldn't nail it down. Maybe I am paranoid with Raya having sole custody of Kaley. Maybe I am so bitter, I am looking for things that are not there?"

"If you sensed something was there, I'd stay with that." Deklan admonished Behr's apprehension. "We are both fathers and I know when something is not right around my boys. I may not always react the way I am supposed to react, but don't tell me I do not know my boys."

"Something was and is wrong in Colorado." Behr was certain, yet he paused during his response to Novak. Behr stared ahead briefly, then clasped his hands together and rested his forehead on his hands while he looked down at the table again. Suddenly, Behr slammed his fist onto the table and looked up at Deklan Novak. The two coffee cups on their table rattled noticeably. The server nearby was startled. Customers seated in the immediate vicinity peered over at the loud interruption. Behr's eyes were cold and distant, somehow bridging the gap between fury and sadness. Deklan stood up to calm the room. Deklan flattened his hands out, while slowly pushing them up and down, gesturing for everyone to relax.

"Nothing wrong here, folks." Deklan spoke calmly but loud enough to be heard clearly. "My friend is going through a rough divorce and got too animated. Our apologies for sure." Deklan sat back down. Behr was embarrassed and hoped no one from the University's administration was having breakfast at Walker Bros. that morning.

"Kaley didn't seem to know who I was, Deklan. Imagine that. Imagine your own son looking away from you when you came 1800 miles to see him for the first time in a year. Am I supposed to let Kaley hate me because Raya has spent the last year implanting that tree? Am I supposed to give Kaley time to allow her to adjust to the divorce? I didn't want the fucking divorce to begin with! If I wait, if I give Kaley time to understand the split between her parents, then I'll lose my daughter for good. I can't do that, Deklan." Behr's angry eyes looked away from Deklan. Northwestern University Defensive Co-Ordinator Behr Thomas had no desire to cry in front of his second-year, bad-ass Navy SEAL linebacker.

"Don't lose her, coach." Deklan announced simply. "What do you know?"

"I know that Kaley never got sick when we were married." Behr began. The spectators and their attention inside the restaurant had returned to their own tables. "Maybe the divorce made her sick? Maybe she blamed me for that? I know that Raya was home-schooling

Kaley for now. Raya said the public school in Steamboat was well below national standards in addition to being in a dangerous section of the town and Kaley was too vulnerable to send her to school this year. Raya also said that they were waiting until I could afford a private school in Steamboat for Kaley and the home-schooling made sense for now."

"Does that make sense to you?" Deklan knew that it did not.

"No, it doesn't." Behr's anger began to rise again. "How dangerous can Steamboat Springs be? Why keep Kaley home if both adults in the home are working during the day? I had never heard anything about a poor public school in a bad part of town before. There wasn't one request for private school tuition until I showed up unannounced in Steamboat. Why not ask me for money from day one?"

"Was Kaley home when you stopped by?"

"I don't know." Behr was embarrassed to think he walked away without pushing the matter more with Gallardo. Behr imagined Deklan would have chosen a different path. "Raya was working. If Kaley was home, I didn't see her."

"Are both adults working during the day, normally?"

"Raya is working at a restaurant in downtown Steamboat Springs. I'm not sure what her hours are. She told me once that she is working both lunches and dinners, but the hours available were not great. I am supposed to know what Raya is making monthly. In the divorce agreement, co-parents with minor children are supposed to be transparent and organized with the finances related to the child's care. What do the parents make in salaries? How much money is being put aside for the child? The clearer the parents know what they can afford, the better things are for the child. I don't know what Gallardo does. Raya told me that he works for the Steamboat Rodeo grounds as their handyman. The finances from Steamboat are unknown. I don't know what Raya makes and I don't know what Gallardo makes or if he is working at all."

"Do you still have life insurance with Raya as the beneficiary or have you changed the beneficiary to Kaley?" Deklan asked because he had researched what to do when and if his divorce happens.

CHAPTER EIGHTEEN

"I haven't changed anything regarding my life insurance policy. The judge at our divorce made a point of telling me that as a part of the divorce, I had to continue to carry life insurance with Raya as the beneficiary until Kaley was not a minor any longer. I have a great package with the team, so I left it intact."

"Check with your insurance guy and find out the exact details regarding the policy awards in the event of your death and who the money can go to. Find out if there is a cash value to the policy and if Raya can access the cash from the policy. Make certain that Gallardo cannot access a dime, even if they get married." Deklan instructed his defensive coach on a common sense move that was often overlooked in a heated and emotional split.

Behr listened carefully while second-guessing virtually every aspect of his post-marriage parenting prowess. Deklan wondered if his marriage was going to end with the same outcome that Behr Thomas had been handed in a divorce courtroom. Most divorce proceedings in the United States are initiated by wives. As a result, many decent men find themselves divorced against their will in a no-fault divorce legal system. Women receive full custody of the children, or the child, as with Behr Thomas and the men are assigned the task of funding the ex-wife's voluntary new single lifestyle.

"Don't beat yourself up, Coach. Somebody had to point this crap out to me, too. When Amber took the boys back to her mom's, I had to talk to an attorney in case we go down the same road you guys did." Deklan sensed the room and lied. "Do you continue to hold any joint accounts for investments or college funding? I'm guessing you don't but call your insurance guy for sure. Again, make sure Gallardo cannot access a dime. What else coach? Should I go with you to Steamboat for any other new recruits?" Deklan smiled trying to ease some of the tension. Behr Thomas smiled and wanted to say, hell yes! "What's good here? I've heard about the place, but haven't eaten here, yet."

"Everything for breakfast is good here." Behr responded. "The omelets look like they are on steroids. The bacon is insanely good, and that Apple Pancake thing is like having apple pie for breakfast. If you want to beef up for the D-Line next year, come here and eat the Apple Pancake all the time." Behr paused.

"I guess I'll have to try the Apple Pancake." Deklan looked for their server.

"You asked me what else." Behr continued. "I went back to Raya's trailer after Raya and Kaley left my hotel. I knew Raya was taking Kaley to the doctor or that was what she told me. I guess I was hoping to run into Gallardo again. No one was there. A neighbor was spying on me while I was trying to catch a glimpse inside the mobile home. I saw the guy through his window as I made my way behind Raya's mobile home to peek through a window. Most of the windows were covered but I did see something that bothered me. There was a padlock on what appeared to be a closet door or a pantry. The door wasn't very big, and it was next to the front door, almost part of the kitchen. I could see partially from the opposite window, across from the front door. The neighbor next to Raya had pushed his curtains aside and stuck his fat face in the window front and center. I was certain that he was going to call the cops. I wanted him to call the cops. I could tell the cops that my daughter lived there, and I wanted to see her, but my ex-wife's asshole boyfriend prevented that from happening. It was a bad plan to return to Raya's mobile home and a confrontation with the police would have turned a bad plan into an arrest. I decided to leave at that point. I figured there was no upside to an out-of-state peeping Tom explanation to the local law."

"Good plan." Deklan concurred. "There's a Native American proverb that states: **Beware the man that does not speak and the dog that does not bark.** I always took that to mean that we should be cautious of people that are suspiciously quiet or unresponsive. A man who welcomes the opportunity to take you down, has no need to talk himself up. A dog that wants to rip out your throat, has no desire to scare you away by barking. The neighbor was setting you up. What were your thoughts on the padlock?"

"That prick is still dealing, or the closet is his gun safe." Behr guessed. "Gallardo has done time for selling coke. I am guessing that he is not working because he was home at noon on a weekday. I went to the rodeo grounds where Raya said that Gallardo was working. The place is called the Howelson Rodeo Grounds and Brent Romick Arena. I couldn't find anyone to ask. All the winter events in Steamboat appear to be scheduled for the ski area and not the rodeo arena. If Gallardo was working at the rodeo arena in the winter, he would have been working there alone. Maybe he is using Raya's mobile home to store his drugs before he sells whatever he is selling. Maybe Gallardo is a member of a Colorado Militia Group and has automatic weapons stored in the mobile home. Either way, speculation, or fact, I have a terrible feeling about how any of this is in the best interest of my daughter."

CHAPTER NINETEEN

2014...THE OFFICE OF HUMAN RESOURCES NORTHWESTERN UNIVERSITY: BENEFIT ELIGIBILITY

THE OFFICE OF HUMAN RESOURCES FOR NORTHWESTERN UNIVERSITY was located on the 1st floor of an office building at 1800 Sherman Ave. The building sat across the street from the western boundary of the NU campus, less than a block away from the Hilton Orrington Hotel and the main downtown district of Evanston. Behr Thomas had planned to call the office to check on what Deklan mentioned about his life insurance beneficiary and the restrictions or lack of restrictions regarding who was able to access the money. Behr intended to inquire if a death benefit was distributed in a lump sum or if there was a cash value to the policy, could anyone other than Raya access the policy. The office informed Behr that no information like that could be given over the phone. Any inquiries about an employee's benefit package had to begin with vetting the inquiry by presenting the proper physical identification. In other words, Behr was required to show up in person and present proper identification before anyone at Human Resources was allowed to discuss any details associated with an employee's benefit package.

Coach Behr Thomas sat in the small waiting area for the Human

Resources Department. The man at the front desk checked Behr's identification, had Behr sign in, and told Behr that Cynthia Bannon would be with him within 10 to 15 minutes. The man at the front desk hesitated at first, but then spoke up.

"How does the team look for next season?" The young man asked. "You guys started great last year but ran into a rough stretch."

"The team looks great for next year." Behr answered and assumed his identification check revealed his position with the football team. At the Power Five conference schools, most offensive and defensive coordinators for the football programs were full-time employees as coaches and not employed as teachers in addition to their coaching responsibilities. "The early signing period for high school recruits was right before Christmas and Coach Adams and the NU program landed 5 of the top 100 recruits in the country for next season, including the quarterback we had been trying to land."

"Awesome." The desk manager asked nothing further.

Fifteen minutes later, Behr sat across the desk from an attractive woman in her early forties. Cynthia Bannon had retrieved Behr's benefit package before he was called into her office. She crossed her legs and reviewed the documents.

"What can I do for you today, Coach Thomas?"

"Thank you for the short notice appointment." Behr began. "I recently went through a divorce."

"I'm sorry to hear that." Ms. Bannon interrupted.

"Thank you, but I'm sure it was for the best." Behr lied. "A good friend asked me some questions regarding my divorce that I was unable to answer, so that is the reason for my visit today."

"Okay. What do you need to know?"

"I need to know that my ex-wife, Raya Nolan Thomas, is the only person that can access the life insurance policy in any form. That means, if I pass and a death benefit is paid, it can only be paid to Raya. If there is a cash value to the policy, who can access the cash value? I want to make sure that Raya cannot take the cash value in the policy and therefore reduce the death benefit of the policy. My ex-wife has a new live-in boyfriend who is bad news. I do not want him to be able to profit from my policy."

"The policy is protected from what you fear. Your ex-wife cannot redeem the cash value of the policy. If you pass and a death benefit is paid out, the only person it can be paid to is your ex-wife. What she does with the money after that is not regulated. No one can dictate to your ex-wife that she cannot share the money or bequeath the money she receives to the boyfriend or a new husband. Hopefully, the scenario you fear regarding the live-in boyfriend does not present itself. Hopefully, you, your daughter and your ex-wife are around for decades to follow." Ms. Bannon tried to clear up Behr's concerns.

"My daughter?" Behr looked perplexed. "How did you know I had a daughter?" Behr asked.

"Oh, it's in the records we have for you, Mr. Thomas." Cynthia Bannon explained. "You added Kaley to the health insurance plan when she was born."

"Oh, I'm sorry. I forgot about that." Behr was embarrassed.

"We also added disability insurance for you once the baby was born." Ms Bannon added. "I also see that you recently took out a child life insurance policy for $200,000 on Kaley Thomas."

CHAPTER TWENTY

2014...A SECOND VISIT TO STEAMBOAT SPRINGS AND THE 14TH JUDICIAL DISTRICT ATTORNEY'S OFFICE FOR GRAND, ROUTT, AND MOFFAT COUNTIES IN COLORADO, MATT PINKLETON 14TH JUDICIAL DISTRICT ATTORNEY

"In each case an assessment is made as to whether a case can and should be prosecuted, and if so, what that prosecution should look like. Some cases call for the aggressive pursuit of a substantial punishment, and others call for a more rehabilitative and compassionate approach. Each case is different, and in each case, we strive to strike the right balance between public safety and human decency. The thoughtful, level-headed, pragmatic application of the rule of law is our daily mission."

-Matt Pinkleton, 14th Judicial District Attorney 2014

The Routt County Courthouse, located at 522 Lincoln Ave. in Steamboat Springs, celebrated its 90th anniversary in 2014. The Oak Creek Times in 1922 noted that the courthouse had withstood an earthquake, although historians of the region found the details and dates of the referenced sizeable earthquake to be elusive at best and folklore more than likely. Whether the building had ever survived a major earthquake was not debated in Northwest Colorado. Matt Pinkleton looked out from his office at the Yampa River and wondered if he would have survived during the era when the District Attorney's offices were built. Pinkleton suspected that he would not have fared well. Matt Pinkleton waited for his morning appointment with Behr Thomas, a football coach from Northwestern University in town to visit his daughter. Pinkleton was told that Mr. Thomas had some legal concerns regarding the welfare of a child. After a suggestion from the DA's office to speak to Child Protective Services, Mr. Thomas insisted on meeting with the Routt County District Attorney on the advice of his counsel, Dwayne Flynn, the head of Northwestern University's Office of the General Counsel. Pinkleton agreed to meet with Behr Thomas.

Spring ball 2014 in Evanston was competitive. The April games involved returning starters and roster players, newly transferred players who were destined to red shirt for a year, and many freshman players who began college in January. Spring practices covered a 34-day period with 20 hours per week of unrestricted activities. Players had to have at least one day per week off. The NCAA stipulated that only 15 on-field practices were allowed, with no more than 12 practices that involved contact. Full contact was not permitted until the third practice. Eight of the 12 contact practices were allowed to involve tackling, and no more than three of the eight could be devoted to 11 on 11 scrimmages. During noncontact practice sessions, headgear was advised. Lance Adams often wondered how much money and time was spent hammering out those rivetingly insightful guidelines.

Behr Thomas was anxious to get out of Dodge as soon as the spring practice schedule had been completed. Normally, Behr was laser focused on the new season, the new roster, the new players, the physical conditioning regimen that his players had been following, and the upcoming Big Ten schedule. Behr was not as focused as he had been during previous years. He planned a trip back to Steamboat

CHAPTER TWENTY

Springs that was going to take place on the Monday after spring ball ended. Deklan Novak was coming to Steamboat Springs with Behr Thomas. After Behr's discovery of the life insurance policy taken out on Kaley; the game had changed. Behr was advised not to query his ex-wife about the policy. Behr met with an attorney at Northwestern University and together they decided Behr should reach out to Matt Pinkleton, 14th Judicial District Attorney for Routt County that included Steamboat Springs, Colorado. The local NU attorney cautioned Behr Thomas to avoid acting emotionally and prematurely regarding the information that Coach Thomas provided to the DA's office. Behr's suspicions to date contained very little hard evidence and the animosity brought forward by a disgruntled, divorced father who had his daughter taken away to another state was destined to be a subject addressed by the local D.A.

Behr Thomas presented the following concerns to the Office of the General Counsel at the Rebecca Crown Center at 633 Clark Street on the Evanston campus, before contacting any authorities in Colorado. Northwestern University General Counsel Dwayne Flynn was happy to assist the Northwestern University Defensive Coordinator. The list of potentially harmful actions surrounding Kaley Thomas was sent to Dwayne Flynn prior to their first meeting. General Counsel Dwayne Flynn read the following list:

1. Raya Nolan Thomas had taken out an unusually large child-life insurance policy and forged Behr's name on the policy. It was unclear who signed the policy, but Behr had no prior knowledge of the policy until he was told by Human Resources at Northwestern.
2. Raya's new live-in boyfriend, Carmen Gallardo, with a felony criminal record, including two prison terms, prevented Behr Thomas from seeing his daughter and prevented Behr from accessing Kaley's home.
3. Kaley Thomas has been in and out of hospitals since the divorce and the move to Steamboat Springs, Colorado. Crucial facts surrounding Kaley's health were hidden from

Behr Thomas, such as more than a dozen trips to the emergency room in Steamboat Springs, Edwards, and Avon. Kaley had 22 medical procedures in Colorado, stemming from multiple infections, breathing disorders, severe vomiting, a loss of appetite to the point of malnutrition, and numerous occurrences of fit-like episodes. Often, the examinations found nothing but a normal child. Other times, the doctors diagnosed fructose intolerance, reflux (back flow of acid from the stomach to the esophagus), sinusitis or several intestinal infections. Kaley had been taken to more than one psychiatrist.

4. Raya Nolan Thomas and Gallardo have held Kaley out from public school and claim they are home-schooling Kaley, a responsibility that does not appear possible if both adults are working full-time, as claimed. When asked about a curriculum for Kaley, Raya refused to produce a plan to educate Kaley.

5. Raya Nolan Thomas has repeatedly refused to verify her income and source of income related to the care of their daughter, a requirement within the divorce settlement agreement.

Coach Behr Thomas and Dwayne Flynn met two days after Flynn received Behr's list. Attorney Dwayne Flynn, a seasoned well-paid attorney and former law professor, headed up Northwestern University's Office of General Counsel as Vice-President and General Counsel. Dwayne Flynn gave Coach Behr Thomas the same advice that the DA in Steamboat Springs was about to give Coach Thomas.

"Coach Thomas." Dwayne Flynn began. "There is nothing actionable in the information that you have provided me. First, you do not have access to Raya's home, where Kaley lives. Whoever decided not to allow you in the home was well within their rights as a resident of the home. Don't get me wrong. There is information that is of great concern, and we will proceed carefully to ascertain what exactly is going on. Colorado Child Protective Services must be notified of

CHAPTER TWENTY

neglect or abuse to schedule an unscheduled home visit. CPS is most often the gateway to secure a child abuse indictment and/or a request to change the custody agreement within a divorce. Child Protective Services in most states have lengthy laundry lists of restrictions that are in place to protect families. Parental protection laws are in place because over 60% of all child abuse accusations reported in this country are proven false. Many marital residual vendettas are carried out by a disgruntled ex-partner through the claims of child abuse. Make no mistake here, Behr, what you are insinuating is child abuse." Flynn paused for the statement to resonate.

"There are laws regarding the homeschooling of a child in each state that are often ignored by parents. Colorado's homeschool statute requires parents to notify the local school district 14 days prior to the start of the school year that the child will be homeschooled. The state requires annual notification thereafter. Parents of homeschooled children are required to follow a universal curriculum related to the grade of the child. Homeschooled children are required to be immunized. The home-schooled child must pass national standardized achievement tests at each grade level through 12th grade. If the child fails to pass the standardized tests, the parents are required to enroll the child in public school or private school. Enforcing the homeschooling statute can be difficult, as it is often difficult to monitor the quality of the education that children receive at home." Dwayne Flynn hesitated long enough for Behr to grasp the information provided by the NU attorney.

"Forging your signature on a life insurance policy is insurance fraud that can result in criminal and civil penalties." Flynn continued. "The difficulty in proving a forgery in court depends on many aspects of the case. If the forgery is professional or exceptional, it can be impossible to prove. There are associated laws that pertain to your wife's "insurable interests" and how she may claim those interests will pertain to your claim. The State of Colorado may take jurisdiction over the case because the fraud, if fraud can be established, was committed in Colorado, where Raya lives. The federal government may take jurisdiction over the case if the fraud was committed across state lines, as it appears. Raya and her boyfriend may face federal and local charges if the fraud involved the boundaries specified by interstate commerce violations."

"You are losing me." Behr interrupted.

"I'm sorry." Dwayne Flynn smiled. "I cannot act on your behalf unless the university is involved somehow, which from what you have told me, is not the case. I can advise you, however. Contact the District Attorney's office in Steamboat Springs, Colorado. Before you contact the DA in Steamboat Springs, make sure you have everything to present him. The DA will want actionable evidence to move forward. Any DA will tell you that while your concern is warranted, there is little to pursue a case except the forged signature and that alone is not enough to risk the case you appear to be presenting. If you suspect your daughter is in danger or imminent danger, then you need more hard evidence to allow Child Protective Services to get a visit on the books and move to your goal of pulling the child from the home immediately before something tragic can occur."

"Nothing I have shown you is enough to get Kaley taken from Raya's care?" Behr asked but knew the answer coming.

"Not as you have presented the case to me." Dwayne Flynn replied. "Behr, there is one thing that stands out to me in what you have presented to me that is more than troubling."

Behr stared at the attorney and waited.

"You told me that your daughter has been in and out of hospitals since she relocated to Colorado and the cause of her health issues has not been categorically diagnosed. Kaley has been taken to the ER on 12 occasions since she moved to Colorado and has had 22 medical procedures during the same time frame? Is that correct?"

"Yes." Behr answered.

"Before you meet with the District Attorney in Steamboat Springs or Routt County, find out everything you can about why Kaley has been sick so much since she moved to Colorado. Any action to reverse a custody decision by the courts will be expedited due to a negative change in a child's well-being." Flynn advised.

"Do you think Raya and her boyfriend are neglecting my daughter

CHAPTER TWENTY 215

and therefore a contributing factor in Kaley's declining health issues?" Behr asked the obvious.

"That is your case, Coach." Flynn replied confidently. "Homeschooling and insurance fraud will tie you up in court for years. The only way to move the needle quickly in a child abuse case is to prove the child is in jeopardy. I am not an investigator, but it doesn't take Hieronymus Bosch to decipher something nefarious going on with Gallardo and Raya that is a contributing factor to Kaley's declining health. Good luck, Behr. If you need to use my office for any support tools, please reach out to me anytime. We cannot officially represent you, but I'll stretch that to my absolute limit."

"Thank you, Dwayne. "Behr stood up and the two men shook hands and then Behr pulled the well-dressed attorney in for a hug. Dwayne Flynn did not resist. Behr Thomas headed for the door.

"One thing you can investigate before you contact the authorities in Colorado. "Dwayne Flynn added.

"What's that?" Behr asked.

"The health issues with your daughter are obviously troubling, but they may bring to the forefront other issues that are rarely examined. We examined a serial killer from England at the NU Pritzker School of Law. I taught there some years back. Beverly Allitt received thirteen life sentences in 1993. Allitt worked as a nurse in a children's ward at two hospitals in England where the crimes were committed. Four of Allitt's victims died. Allitt was charged with four counts of murder, eleven counts of attempted murder and eleven counts of causing grievous bodily harm. Allitt's motives were never fully explained, but many believed she showed symptoms of a psychological disorder. The disorder is called factitious disorder imposed on another (FDIA), previously or first named Munchausen syndrome by proxy (MSbP) back in the late seventies. The disorder is a condition in which the caregiver creates the appearance of health problems in another person, typically their child. In Allitt's case, the victims were children but not her own. The behavior is child abuse, difficult to detect and much more difficult to prosecute." Dwayne Flynn wrote down the clinical names for the condition he mentioned and handed the paper to Coach Thomas.

"Talk to the medical staff at the facilities where Kaley was treated and ask about FDIA before you talk to the DA in Steamboat Springs." Dwayne Flynn could not have been more helpful. Dwayne and Behr Thomas then stood silent briefly as Behr reviewed the notes. Behr nodded and left the office.

Behr Thomas and Deklan Novak flew to Yampa Valley Regional Airport during the last week of April, after the spring football games had been held at Northwestern. Yampa Valley Regional Airport served Steamboat Springs, Hayden and Craig, Colorado inside Routt County in Northwest Colorado. The airport was 25 miles west of Steamboat Springs and 2 miles southeast of Hayden. Behr and Deklan took a Southwest flight that stopped in Denver where they changed planes. The pair left at 7:30 a.m. CST from Chicago's O'Hare International Airport and arrived in Steamboat at 1:30 p.m. MST.

During the plane ride to Steamboat Springs, Deklan sat next to Behr Thomas and Deklan came to a new epiphany that he hadn't foreseen as a crossroad so soon after leaving the Teams. More than 18 months before the trip to Colorado, Deklan had said farewell to the joint missions where United States Special Forces were routinely ordered to team up with inexperienced Allied counterparts that had life and death consequences. Deklan's past year and a half had been spent playing football and struggling to maintain a respectable grade-point average at Northwestern as a full-time student. Deklan's marriage was unraveling and the Afghan Army's inability to keep up with Navy SEAL led missions had not been a thought process since Deklan arrived at Great Lakes Recruit Training Command for his last stop in the Navy. Deklan helped run the pre-BUD/S intro class at Navy boot camp for aspiring Navy SEALs. Now, on the plane to Steamboat Springs, Deklan realized what a liability Behr Thomas was about to become.

"I want you to stay at the hotel when I go to visit Raya's trailer park home. "Deklan announced out of the blue. "The original plan where we must enter the trailer when no one is home can go wrong at

CHAPTER TWENTY

least two-dozen ways. In the Teams, when we planned our assault missions, we planned for many contingencies that were very unlikely to happen, but the planning was the key to success. When all the planning was complete and the gear and weapons were checked and re-checked, we set out on the op. What was always in the forefront of a good operator's mind was that no matter how well we had prepared and planned for the unexpected, we most likely had only covered about 50% of what could happen in combat. The best plans are often shot to shit in a New York minute and then all hell breaks loose. The SEAL Team Six guys that took down OBL, lost a chopper before they had set down on the ground at the Abbottabad compound. The planning and execution of the OBL mission was completely changed before the first combat boot hit the ground. The mission was designed for and planned for two choppers and 24 Team guys." Deklan suddenly missed his former occupation.

"The Team practiced the mission repeatedly with two stealth Black Hawk helicopters and never got the green light for Operation Neptune Spear until SEAL Team Six proved to the President, the CIA Director, the head of JSOC (Joint Special Operations Command), the Secretary of State, Defense Secretary, the Chairman of the Joints Chiefs of Staff, and a host of other National Security Advisors that they could take down OBL with only two choppers and 24 Team guys. Before the mission engaged, they had lost half of their exfil capabilities and had incurred an unknown number of casualties. As it played out, there were no American casualties. The first bird to land at the target in Abbottabad, was supposed to land on the third level of the compound and the second bird was supposed to land inside the first-floor compound walls. The mission had planned and trained extensively for a penetration assault where the attacking forces came from above and from below the target. In the Abbottabad compound, the target was the third-floor residence of OBL. All the planning was based on a surprise attack with little time for the sleeping enemy to prepare for the incoming forces. The Team guys landing on the third level would attack from the top level. The Team landing inside the compound walls on the ground would lead the assault up through the compound. They had planned to meet in the middle somewhere. The first bird crashed outside the compound and the second bird landed well outside the compound to assist at the crash site and extract the team from the downed chopper. The entire ordeal set back the mission maybe two minutes but woke up the entire compound and gave everyone a chance inside the compound to grab a weapon and prepare for the assault. Mike Tyson's adage lived out in real time back in 2011..." ***everybody has a plan until they get punched in the face.***"

"I'm sorry Deklan." Behr interrupted. "We are taking on this task to get my six-year-old daughter out of harm's way. You are not going anywhere without me, Rambo. I don't care how bad things go wrong. Can we get past the "you stay at the hotel" bullshit and deal with me and my inexperience on the mission from now on.?" Behr Thomas was adamant. "So, help me God, Deklan, I am not staying at the hotel no matter what you tell me."

"I figured you'd say that." Deklan smiled, but the smile didn't linger. "You and I will do this thing, but you will listen to me on everything we do and what I say goes no matter what we encounter. Are you clear on that, Coach?"

"I wouldn't have it any other way." Behr Thomas concurred.

"By the way, Coach. "Deklan added dispassionately. "God will have little to do with the success or failure of our goals to bring Kaley back home. We will fix the problems your daughter is confronting, and we will eventually reunite Kaley with you permanently, but neither you nor I will depend on anything or anyone other than ourselves. I will dictate what we do here, where we go here, and when we pull the trigger on every action. Back on the field when this is over, you rule. Here, I am the Special Warfare Commander…one rule, one word.

"Agreed." Behr repeated and felt secure, a bit relieved and was very glad Deklan Novak was not looking for him.

They checked into the Hotel Steamboat Springs and took in the unpleasant building odor as they got their room keys at the front desk. The hotel didn't smell from gas leaks, plumbing issues, uncovered trash bins, or mold. The hotel smelled like an ashtray, as it did on

CHAPTER TWENTY

Behr's first visit. The lingering cigarette scent hung inside the core of the hotel like Sean Connery's Scottish brogue from a James Bond film.

The April ski season was busy in Steamboat Springs with late spring skiers taking advantage of the massive winter snow volume and the lingering cold weather. There was a cacophony of skier echoes inside and around the hotel, highlighted by the rattle of newly arriving snowboards, skis, poles, and boots dropped at the main entrance with the bell boys. The visitors pulled their luggage as the wheels rattled on the ceramic tile floor and the rhythmic clattering bounced off the stone fireplace, warm wooden walls, high-beamed ceiling, and the frozen animal heads hanging from the walls like an acoustic echo chamber. Behr and Deklan carried back-packs and a single carry-on bag each. They were not in Steamboat to go skiing.

Behr and Deklan rented a 2012 Ford E-250 cargo van (E-Series) with an 8-cylinder engine, automatic transmission and front-wheel drive. Deklan came up with the penetration and exfil plan for their trip to Steamboat Springs. The original plan called for complete radio silence between Behr and Deklan, a series of at least a dozen drop-dead times and contingencies surrounding each drop-dead time sequence. Drop dead times and their alternatives were a crucial element to tactical planning outlined in the Commander's Tactical Handbook and the Logistics Handbook for Strategic Mobility Planning. Each drop-dead time deadline had alternative parameters once the timeline was compromised. Deklan and Behr left their cell phones back in Illinois. Deklan had a former SEAL teammate that was still stationed at Great Lakes take both cell phones back to his apartment. The phones could not sit idle on a countertop for four days if the phones were ever to be used to verify an alibi for where Deklan and Behr had been during their Steamboat Springs excursion. The phones would need to be used at various times during their trip and at various locations…like a Home Depot, a McDonald's location, or at a Walker Brothers Restaurant in Wilmette, a favorite eating spot for both men.

Before they left, Deklan had ordered two 12" x 18" magnetic signs from an internet sign company that they carried with them to Steamboat Springs. The Centurylink logo ran horizontal and the green circular mosaic design next to the company name was visible on both sides of the van. Behr and Deklan purchased white construction helmets and Dib certified ANSI (American National Standard Institute) Safety Reflective Yellow Mesh Vests for high visibility. The white hats and mesh vests were what Centurylink servicemen wore on home installation and repair calls. Behr would wear the Centurylink garb. Deklan would not but kept a Centurylink uniform if needed.

Deklan's plan involved at least 2-3 days of surveillance to determine what time during the night or day was best for insertion. Deklan and Behr were in Steamboat Springs to gather evidence that Kaley was being abused for the sole purpose of Behr Thomas regaining custody of his daughter. Any felony charges leveled against Raya and Gallardo would be desired but gathering actionable evidence to bring Kaley back to Chicago as quickly as the courts allowed was the sole goal of the trip.

The pair studied the health angle presented to Behr by the attorney Dwayne Flynn. Behr imagined that they might be able to find evidence of the medical fraud that Flynn alluded to. Deklan had to look for Lysol bottles or any kind of poison. What prescription drugs or firearms were in the mobile home and did the circumstances present imminent danger for the child, consistent with the state's definition of reckless child endangerment? What else would they find in the trailer? Behr was convinced they were about to uncover a drug distribution network being run out of the trailer home park by Gallardo. Behr was certain they would find a plethora of illegal weapons and drugs for distribution that any court in America would find unacceptable and be enough to reverse a court-awarded custody decision. The locked cabinet inside the mobile home was an indelible image that Behr could not shake. Behr was convinced the cabinet held the guns and drugs. Deklan and Behr assumed they would find more than enough evidence in the mobile home to change Kaley's address. The weapons and drugs needed to be photographed inside the mobile home and established as actionable evidence for the police or Colorado Child Protection Services to find obvious reckless child

CHAPTER TWENTY

endangerment, actions punished as a felony if the parent or guardian's environment placed the child in harm's way. Everything about an unattended firearm in a child's home placed the child in harm's way.

The plan consisted of two surveillance objectives. First, were the incriminating photographs sufficient to expose the abuse and illegal activity. Next, Deklan planned to install a hidden video camera to record Gallardo and Raya inside their home. The video feed from the camera would be recorded on Deklan's laptop. It is legal in Colorado to videotape someone in the home without their knowledge. It is illegal in fifteen states to include audio without someone's consent. Colorado was not one of those states. The video evidence would then be brought to the FBI or local police authorities so they could get a search warrant and legally prosecute the couple, while the courts would take immediate action to return the girl to her father. The Steamboat Springs Police Department had 12 hours to report suspected child abuse or neglect to Child Protective Services and the county ran a 24-hour CPS emergency response hotline. Evidence obtained from trespassing in Colorado was admissible in court if the evidence confirmed a crime against a child. Depending on the judge, the plan was not perfect. Most were not.

Deklan Novak was back in Special Forces. The past two years, he had not been required to plan and re-plan missions that could result in his demise and the deaths of his SEAL brothers. The plan that Behr and Deklan were rehearsing and the success or failure of the plan, ultimately would determine the safety and survival of a six-year-old girl. What if things went south during the operation? Deklan was used to planning missions with inexperienced partners. In fact, most of the Allied friendlies in Iraq and Afghanistan only hindered the SEAL Teams when missions were joint efforts. The SEAL Teams had their hands tied in so many ways during the war on terror. Dragging untrained partners on dangerous missions was common.

The details of the trip began to unfold. Behr was to stay with the van, dressed in a Centurlink-like uniform and driving the rented white van with the company logo on the side. A dormant Centurylink van waiting by an electrical transmission tower would not garner any unusual attention from a passing local law enforcement vehicle or a park ranger. The van was the extraction ride once Deklan left the mobile home park and made his way to the van through the mountain trails behind the mobile home park. Deklan had to be certain his exit was clean, before he hooked up with Behr and the van. The insertion point for Deklan would start from above the mobile home park on Emerald Mountain. The nearby ski area was busy and that helped to deviate from any potential alarm associated with the insertion point. Cars, trucks, and vans were all over the area. Howelsen Hill Ski Area had produced 89 Olympians in both alpine and Nordic events since 1931. Howelsen Hill Ski Area was the oldest operating ski area in North America. The grounds included a series of ski jumps, which separated Howelsen Hill from most alpine ski centers. The plans had to be memorized by Behr Thomas. No notes would be written down. If either man was ever stopped and/or questioned, there could not be a paper trail listing the drop-dead time parameters.

Deklan's plan was based on the theories of death from above. Deklan would drop in from higher ground. The exfil routes would have to be run repeatedly. Deklan Novak was not going to be caught by an inexperienced team of local law enforcement, no matter what went wrong with the plan. Deklan was extensively trained in breaching any kind of a compound, including the fulfillment of a surreptitious entry, a non-destructive method of entry that would be both undetectable during normal use and undetectable during an inspection by law enforcement. Any incriminating photographs had to be obtained without breaking into the mobile home or at least, appear to be obtained by an approved entry. If there were no visible signs of a forced entry, there were any number of reasons that could exist whereby Behr could claim how he obtained the photographs. Raya had invited him inside the mobile home. After all, Behr's daughter lived at the location. If Raya told the authorities she did not allow Behr to enter the home, she would have to come up with a reason. Behr's concern for Kaley's health over the past year was a sufficient reason to visit the home. Regardless, the evidence would

CHAPTER TWENTY

have been presented and a CPS visit would be warranted regardless of any legal ping-pong.

In the event of a plan going awry, which often was the case, Deklan had planned for a couple extraction diversions. Plans go wrong more times than not. The plan was to enter and exit the mobile home when no one was in the trailer. The main deterrent to entering the mobile home when Raya was working an evening shift at the Eighth Street Steakhouse and Gallardo was holding court at any of several local watering holes, was Kaley. Where was Kaley? Did Raya and Gallardo leave Kaley alone at night. If this was the case, Deklan knew they had game, set, and match. Leaving a six-year-old girl to fend for herself is not looked upon favorably by CPS. Most states considered leaving a six-year-old child alone for four hours or more to be actionable child neglect. If Kaley was home, the plan would have to take a drastic turn. If Kaley was not home, where was she? When Behr had gone to the mobile home twice, he had not seen Kaley. Was she hospitalized? Was she with a neighbor, who was hired to watch Kaley when Raya was working? Gallardo did not hang out much at the mobile home when Raya went to work. Deklan and Behr's surveillance revealed the daily patterns followed by Raya and Gallardo, but Kaley had not been seen by either man during the entire surveillance period. Deklan thought the worst but did not discuss his concern with Behr. Deklan wanted to get inside the mobile home before any nefarious theories were presented. Behr Thomas imagined much the same without any discussion.

If Deklan was discovered at the mobile home during his insertion, there were two plans in place that Deklan and Behr rehearsed. If the insertion was clean and undetected, Deklan would exfil through the mountain paths next to Howelsen Hill and meet up with Behr and his Centurylink truck at a nearby rendezvous location on one of the access roads leading down from the mountain. If the insertion was compromised, Deklan carried a thermite grenade, homemade and built to start a fire but not built to blow up an enemy compound. Thermite grenades were a pyrotechnic composition of metal powder and metal oxide. When ignited by heat or a chemical reaction, thermite undergoes an exothermic reduction oxidation and will create a brief burst of heat and high temperatures in a small area. If detected during the break-in, Deklan decided that he would divert the attention to another mobile home. Deklan would start a fire at a neighbor's mobile home, leaving a note on the targeted residence that would lead law enforcement to believe the target of the blaze was the neighbor's mobile home. Law enforcement would believe the blaze was brought on by an enemy of the neighbor attempting to get even for some made up circumstances detailed in the note. The note that Deklan wrote up with a black marker on a discarded, torn ski magazine cover stated:

"*I tried reasoning with you for months. I asked you to make sure that dog was tied up or inside. I told you that ignoring me had consequences and now you know I was not full of shit. Eat shit asshole! Don't ever threaten my family again with that fucking dog!*"

The note was to be destroyed if the exfil was clean and undetected. Deklan would eat the note. The blaze would have to be small enough to be contained outside of one mobile home, in or near a vehicle, without setting the home on fire. With Deklan's training, he could set a go-cart on fire inside a garage filled with full gasoline cans and not damage the cars in the garage or ignite the gasoline. Deklan was not in Steamboat Springs to kill anyone, so Deklan and Behr had to determine a nearby mobile home to Raya's where the diversion grenade was to be ignited. The mobile home had to have been home to a large dog or a pit bull type of pet. Surveillance revealed a couple of potential target homes. The surveillance days would be used to watch for the right dog and pinpoint the acceptable mobile home for a potential diversion.

The surveillance days were long and tedious but necessary. Both Behr and Deklan drove around in the white van and wore the CenturyLink uniforms. After watching Raya's mobile home for three days and following Raya and Carmen Gallardo for three days, the pattern emerged and proved operational. Raya worked the dinner shift in town. Raya left the mobile home just before 4:00 to be at work on time and ready for a 4:30 p.m. start to the dinner shift. Gallardo left shortly after Raya left, if he was home at all. Gallardo hung out at

CHAPTER TWENTY

mainly two bars in Steamboat Springs. He hung out at Sunpies and Carl's Tavern just down the street from Sunpies. Gallardo and Raya didn't venture out much during the day. There was a trip to the grocery store once and Gallardo went to a local body shop once and then stopped at a liquor store on the way back to the mobile home. Both Raya and Gallardo were back at the mobile home park between 10:00 p.m. and 11:00 p.m. on each of the three days Novak and Behr Thomas played private detectives.

The plan emerged for Deklan to visit the mobile home after the sun went down sometime after 7:30 p.m. Deklan decided to break into the mobile home at 9 p.m. The front door had been reinforced with a security plate below the door's window. There was an added dead bolt lock to the front door. Deklan could get past both obstacles in less than a minute. Behr and Deklan wondered what was going on inside the residence that inspired Gallardo to reinforce the front door and to add another dead bolt on the same door. Security and safety for a young girl and her mother, Deklan concluded, was not high on the list of possible reasons for the added security. The biggest mystery that had yet to be revealed was the whereabouts of Kaley. During a three-day surveillance, Deklan and Behr never saw Kaley. Was she in the mobile home? Was she in a hospital? Raya and Gallardo never went to a hospital during the three-day surveillance. Was Kaley not allowed outside? Was Kaley a prisoner in the home? Deklan Novak was a seasoned Special Forces operator and not prone to catastrophizing. However, the obvious had to be considered. Was Kaley dead?

CHAPTER TWENTY-ONE

SPRING 2014...THE BREACH AT THE DREAM ISLAND MOBILE HOME PARK

THE PRACTICE RUNS were difficult and necessary. There were a few marked country roads behind Emerald Mountain near the quarry. Aspen Street led down to Fern Hill Road and then to Scrub Oak Path. The white van dropped Deklan off behind Emerald Mountain on Country Road #36. Behr drove back to the hotel during the practice runs and headed to Scrub Oak Path for the pick-up at the agreed upon time. Deklan insisted they practice without communication during the runs. Deklan practiced and prepared as if each test run was the real deal, because each test run was the real deal. The terrain was not aware of the practice nature of the trek. The wildlife was not privy to the practice schedule. Deklan always carried a Glock 17, 9 mm pistol with one magazine inside the gun, a round in the chamber and two back-up magazines. Each Glock 17 magazine carried 17 double-stacked staggered rounds. Prior to the trip to Steamboat Springs, Deklan had contacted a former SEAL teammate from SEAL Team One, Pat Gabriel, who lived in the Denver area. Deklan's former teammate delivered a Glock 17 with all the requested ammunition and magazines to Deklan at the hotel in Steamboat. Denver was 156 miles from Steamboat Springs. When the trip to Steamboat had concluded, Deklan would leave the Glock 17 in a wrapped shipping package at the hotel for Pat Gabriel to pick it up at his convenience. Gabriel never said a word about being asked to make two trips to Steamboat Springs and never asked Deklan why he needed a gun. The night vision goggles (NVG), Ground Panoramic Night Vision Goggle (GPNVG-18) from L-3 Warrior Systems and acquired during Deklan's SEAL days, were strapped to Deklan's belt as was the 5" x 7" mountain survival kit or the Navy SEAL Escape and Evasion Kit as the gear was called in the Teams.

The trek from the back of Emerald Mountain and Country Road #36 to the Dream Island Mobile Home Park took roughly 2.5 to 3 hours on foot. Deklan ran the route at night three times and practiced two alternative routes if the first exfil route was compromised. The obvious danger traveling at night was Deklan's unfamiliarity with the terrain. The weather was still cold, hovering in the upper 20's at night in April. The average low temperature in Steamboat Springs in April was 26 F. Aside from the cold and the unknown terrain, Deklan was bound to encounter the Northwestern Colorado wildlife during his climbs down and back up the mountain backcountry behind Steamboat Springs. The Colorado backcountry was home to larger mammals like Bighorn sheep, black bears, mountain lions, coyotes, and Pronghorns. There were hundreds of smaller species of wildlife, but Deklan was not overtly concerned about the gophers, rats, rabbits, beavers, prairie dogs, squirrels, porcupines, and chipmunks.

After what Deklan decided was sufficient practice time, the day came when the breach was to be executed. Deklan and Behr were worried about Kaley but did not discuss their concerns because men are inherently superstitious. Behr knew if he discussed the weather, invariably the weather would turn bad. If the traffic was discussed, invariably the traffic would become horrific instantaneously. Neither Behr nor Deklan wanted to be the cause of something bad happening to Kaley. Deklan got into the van for the ride to the backcountry. Behr was dressed in the Centurylink vest and helmet. Deklan wore Solomon X ultra boots, alpaca wool long socks, and black Arc'teryx alpha waterproof pants. Deklan loved cotton t-shirts, but cotton easily

CHAPTER TWENTY-ONE

retained moisture and on a wintery mountain night, Deklan wanted to stay dry. The UDT/Navy SEAL Instructor shirt was a polyester and rayon blend. Deklan also wore a gray Patagonia knit quarter zip sweater, a black Arc'teryx Macai down insulated Gore-Tex jacket, a Black Mountain Hardware knit beanie hat, and gloves.

Deklan gave a hand signal to Behr and the van disappeared beyond the black night sky, lit up like the astronomical observatory at the Adler Planetarium. Deklan had Behr kill the headlights a half mile from the drop off point. The sky was clear, and the moon gave plenty of light to drive the unknown road. The grade of the incline was not a challenge for Deklan as he hustled up and down the face and backside of Emerald Mountain. During BUD/S, the SEAL candidates carried 70-pound backpacks during ruck marching or rucking up steeper grades for hours. Deklan recalled SCPO Charlie "Cobra" Coletti and the 200 miles run during Hell Week. A 2.5-mile trek with a small fanny pack and a Glock was a stroll in the fresh-cut grass at summer camp in Kenosha. The pine trees rose up off the sloping terrain and painted the moonlit sky with a silent evergreen silhouette landscape. Deklan began to sweat under the clothing he layered for protection and warmth.

The back edge of the mobile home park came into view as the trek took less than three hours. The park was quiet at 8:45 p.m. A dog barked at nothing. Deklan eyed the diversion target as the mutt continued to bark at nothing. Another canine followed suit, and the pair seemed to be talking to each other, oblivious to the intruder clad in black and moving with the stealth of the mountain lions that may have watched him move down the mountain.

Deklan approached Raya's mobile home. The porch was not illuminated. There was a small landing area with a covered entryway. The small light fixture next to the front door was not turned on. The unit was divided into two bedrooms, a living area/kitchen, two tiny bathrooms and a small closet next to the kitchen area. Deklan approached the front of the unit slowly, while keeping his eyes on the unit directly to the right of the front entrance. The beat-up road running next to the mobile home was called Dream Island Plaza.

There were no cars in the driveway next to Raya's mobile home. Deklan waited on the dark porch for about two minutes. There was no movement from inside the unit and no movement from the unit next door to Raya. There was one light on inside the mobile home next door that Deklan could see, but he saw no movement or occupants. A trash filled empty lot stood on the other side of Raya's trailer, A mobile home had been removed from the lot some months back and the footprint was still visible. Deklan studied the front door, the locks and the security plate added to the door. The additions were a waste of money. Deklan picked the lock in less than a minute. Deklan closed the door quietly and slowly. There were no lights on in the unit except for a fluorescent light above the sink. Deklan took out a military tactical LED flashlight.

The mobile home smelled like a dirty commercial kitchen. A distinct level of grease covered each exposed countertop and appliance. Deklan didn't need to touch the surfaces to know the grease was present. Deklan could see the sheen bounce from the fluorescent light. There were dirty plates in the sink and a stack of unwashed dishes on the small counter next to the sink. A small dining table had a half-dozen empty beer cans and some trash from a fast-food restaurant that had not been thrown out. The floor looked like it hadn't seen a mop since it was installed.

Deklan moved the light beam around the kitchen area and the main living area. Deklan was looking for things that belonged to a child. What he saw was more like a frat-house basement after a pledge party. Deklan noticed a half-open kitchen drawer with tin foil and some straws visible. After a closer look, it was an obvious crack stash drawer. Deklan pulled the drawer open all the way and he could see glass pipes or "stems" used for smoking crack. The pipes were hollow glass tubes with bulbous ends that had brown smoke residue and black burn marks on the outside end used for lighting the crack. There were small copper mesh screens in the kitchen drawer and two torch lighters. Deklan took photographs of everything.

On a small kitchen table, Deklan noticed a laptop computer. Deklan unplugged the charging cord from the wall outlet and stuffed

CHAPTER TWENTY-ONE

the laptop computer into a folded backpack that Deklan had stuffed into his Escape and Evasion Kit (E + E kit).

Deklan did not see any clothes, shoes, boots, coats, or toys that would have belonged to a child in the front living area. Deklan walked back to the two bedrooms and wanted to gag. The place was filthy. The tiny bathrooms were disgusting. Wet towels were strewn in a ball on the floor of the first bathroom. A waste basket overflowed with trash. The mirror was cracked and looked like it hadn't been cleaned in a year. A roll of toilet paper sat on the sink. The toilet bowl was stained and cracked. Deklan opened the small medicine cabinet. There was a razor, some make-up containers, a lipstick pouch and two bottles of Ipecac syrup. Ipecac syrup was a stomach-purging product often used to induce vomiting in cases where something toxic had been swallowed. Deklan took photographs of the medicine cabinet, the Ipecac bottles, and the filthy conditions of the unit.

While Deklan was searching the second tiny bathroom and the second bedroom, a noise from the front stopped his movement. Deklan wasn't certain if the sound came from outside the unit or if it came from inside the unit. Deklan's first instinct told him that the sound was from inside the unit. The short sound almost sounded like a child's voice. Deklan's heart froze. Was Kaley in the unit? Deklan moved into the living area and into the kitchen space.

"Hello." Deklan spoke up softly. "Kaley?"

Deklan heard something from the locked cabinet, maybe movement or maybe an inaudible response. Deklan knew at that instance that Kaley was in the cabinet. What he did next had everything to do with assessing the circumstances under duress. Was the cabinet wired? Did Gallardo want Kaley to be found? If so, why?

"Kaley." Deklan moved closer to where Kaley was located and moved to the locked cabinet doors. "I am a friend of your father's, and I am going to open this cabinet, so don't be scared. I am here to help you. If you think of any reason, I should not open the doors, tell me now." Deklan paused and did not make another sound while he waited.

Deklan did not hear a response. Another minute passed.

"Kaley." Deklan spoke calmly and softly again. "I am going to open the cabinet doors now. Don't be afraid. I am here to help you, I promise." Even children can sense the authenticity in a voice.

Kaley was scared to death. Gallardo had warned her about speaking to anyone about what was happening to her and her mother. Deklan pulled out an Ontario MK 3 Navy Knife and jammed the blade into the locked clasp attached to the cabinet doors. Deklan popped the lock off like it was made of plastic. The doors swung open, and Deklan gasped. The cabinet was no more than three feet high and 24" across. Kaley was sitting on a filthy blanket next to a small bucket that she used for a toilet. Kaley was tiny and appeared malnourished. The enclosure smelled like an animal's cage that had not been cleaned in weeks. Kaley held onto a floppy brown teddy bear, stained, and matted. The teddy bear had a big brown nose and was looking down as if the bear was ashamed that he couldn't help Kaley. Deklan thought the bear wanted to talk. Regardless, the little bear told Deklan everything through soulful sad eyes, like small windows into the night. Deklan reached out for Kaley, and she flinched and moved backwards, but Kaley could only go back a couple inches. Deklan waited and did not move quickly.

"I am here to take you to your father. We are here to end this for you. Trust me, Kaley." Deklan held out his hands but made no sudden movements. Deklan had seen babies blown apart in combat zones. Deklan had witnessed unspeakable things done to children that language could not convey. Opening the cabinet doors for Deklan was like getting hit square in the chest by an RPG. Any questions regarding his involvement in Colorado vanished on the spot.

Kaley wanted to cry but wasn't sure what to do. Deklan stayed still and asked, "I don't know much about God Kaley, but I know he brought me here tonight. Let's get out of here and get something to eat. What do you like to eat?"

Kaley moved slowly towards the stranger in her trailer. There was something soothing about the man in black clothes with the bright flashlight. The stranger had kind eyes. Kaley hadn't seen kind eyes in so long. Kaley moved slowly out of the tiny enclosure and into

Deklan's arms. Kaley held her bear tight. The muscular ninja-like figure hugged the little girl like she was made of glass. Deklan didn't speak for what seemed like an eternity. With one hand, Deklan snapped as many pictures as possible of the nightmare that he witnessed and then put his phone away. The ex-Navy SEAL pulled the child tighter into his chest and whispered.

"We are leaving now. I am going to look for your coat first. Do you know where the coats are?" Kaley didn't answer. Deklan wasn't looking for coats. The tiny little girl was buried into the chest of a former Navy SEAL that had come to rescue Cinderella.

Deklan did not ramble or try to confuse Kaley. He held her as he reached behind his back and pulled out a small pouch from his fannypack setup. Deklan took a small, battery-operated video camera, like a nanny-camera and placed it on the ledge that held the fluorescent light over the sink. Deklan activated the camera, had little time to run sight-line checks and had to aim the camera visually to encompass the living area, the locked cabinet, and the front door. Deklan closed the cabinet and reset the lock as best he could. The doors appeared locked, but one touch and the lock would fall to the floor.

Dcklan found another child's blanket in the girl's bedroom and a small coat. Kaley was wearing pajamas and a pair of dirty sneakers. Deklan took out his burner phone and called Behr.

"Get down here to the trailer, now. I've got Kaley and we need to get her out of here before anything else happens. She was locked in the cabinet you thought was holding weapons and drugs. The reason we never saw Kaley during our surveillance was because they had her locked inside a tiny cabinet. Don't exceed the speed limit and don't get out of the truck. When you get here, we'll come out. Hurry Behr but be smart. This is the one chance we are going to have to get Kaley out of this rat hole for good. The video system is in place." Deklan hung up and wrapped Kaley up tight in the blanket he found.

Deklan began scanning the room for anything else to photograph and began taking random shots of everything. Deklan found a 30-06 Winchester shotgun and .308 semi-automatic rifle in the closet inside what had to be Raya and Gallardo's bedroom. Three boxes of shells were behind the guns. Deklan took photographs and left the guns exactly where he found them.

Deklan closed and locked the front door. He took out his knife again and jammed it into the front door lock. If there was any resistance from law enforcement (a certainty) on the legality of finding Kaley, Deklan planned to say they had come to the home to see Kaley, and no one answered the door. They tried again and heard cries for help. With a possible child in danger, they entered the home forcibly and found Kaley locked up. A civilian is allowed to enter a home without permission, or a search warrant if the civilian hears cries for assistance from inside. If law enforcement asked about why the two men didn't call the police immediately after they found the child, Deklan and Behr would explain that they did not know where the occupants of the trailer were and getting the child out of harm's way was the priority. Any call to law enforcement meant waiting for the police to arrive. Behr and Deklan, being strangers to the area, had no idea how long a response from the police might take and opted to remove the child first.

"Your father is on his way to pick us up. He's going to get you whatever you want to eat and then we can go back to his hotel, and you can take a bath and have some ice cream. Does that sound okay with you, Kaley?" Kaley did not say anything. Kaley just held onto her bear and held onto Deklan like she was Velcroed to his chest.

CHAPTER TWENTY-TWO

SPRING 2014...THE STEAMBOAT SPRINGS POLICE DEPARTMENT AND CHILD PROTECTIVE SERVICES MEET KALEY THOMAS

RAYA NOLAN THOMAS and Carmen Gallardo got home around 11:00 p.m. They had left Kaley locked in a kitchen cabinet for what they had assumed was just over 11 hours. In most states, including Colorado, child abuse has been determined to be present when a child six years old or younger is left alone for four hours or more. Of course, in a misdemeanor case, that must assume that the child is left in the home unattended but not locked in a cage or shackled to a bedpost. Child abuse very quickly can move from a misdemeanor to a class 3 felony in a case like Kaley's. Gallardo was intoxicated, having been at one of two bars for most of the night. Gallardo had consumed more than a dozen beers and multiple shots of El Jimador Tequila Blanco, a generic cheap tequila used as a common well-brand. Gallardo waited for Raya to finish her shift at the steakhouse and then he drove them both back to the mobile home park. Raya was not thinking about Kaley or Gallardo's inebriated condition. Raya was thinking about the pipe and smoking as soon as they arrived back home.

Gallardo managed to navigate the short ride without getting stopped for a certain DUI. The beat-up 2006 Toyota Tacoma pickup pulled into the gravel drive next to the trailer home. Raya thrashed her feet around and through the accumulated trash on the floor of the truck as they pulled up in the driveway.

"When are you going to clean this crap out of the truck?" Raya asked as she opened the truck door.

"Right away, honey." Gallardo shot back sarcastically.

"Jesus Christ." Raya mumbled and shook her head, always careful not to go too far with Gallardo.

Gallardo reached the door and froze.

"Stay there!" He shouted. "Someone is inside."

Raya stood still and finally thought about her daughter. Gallardo motioned with his hand for Raya to move to the side of the truck while he was going to enter the trailer. Gallardo was not armed and did not have a gun in the truck. The rifle and the shotgun were in the bedroom and there was a Ruger 38 revolver in a box under the bed. Gallardo pushed the door open slowly. The lock had been pulled apart from the door. The fluorescent light above the sink was still on as Gallardo slowly moved inside. He did not hear anyone in the trailer.

"Kaley?" Gallardo spoke loudly to announce that they were home. There was no answer. The door was pushed open more and the room was entirely visible, but Gallardo waited to enter. He looked around inside as best he could. The cabinet was still locked. The intruder had left or was hiding in a bedroom. Gallardo looked around for a weapon. There was nothing on the front porch and nothing right inside the unit that he could use. There should have been a tire iron in the truck, but Gallardo had no idea where it was. Fuck it, he decided to go in quickly and grab a knife from the kitchen. Carmen Gallardo rushed into the mobile home and rattled around the kitchen until he found a kitchen knife in a utility drawer. He stood silent with the knife, crouched in a fighting stance. Gallardo wouldn't know what to do with a knife in his hand if he was standing in front of a turkey on Thanksgiving. There was no sound.

CHAPTER TWENTY-TWO

"Who is in here?" Gallardo yelled as if his decibel level would send the intruder fleeing or induce instant panic and the intruder might surrender. There was no answer, no panic and no surrender.

Gallardo rushed into the first bedroom and saw no one. Gallardo dropped the kitchen knife and reached under the bed for his revolver. The box was pulled out, opened and Gallardo had a gun. He stepped into the second bedroom and saw no one there. Gallardo walked back out to the main room. The intruder was gone. Raya peaked inside the trailer. Gallardo held out his hand for Raya to wait. He reached for the cabinet and the lock dropped to the floor. The cabinet doors eased open slightly. Gallardo panicked and ripped open the cabinet doors. They slammed against the trailer façade and bounced back and forth. The cabinet was empty. Kaley was gone. Someone broke into their mobile home and took Kaley. Gallardo spun around and yelled.

"Raya!" Gallardo pushed open the front door for Raya. "Your daughter is gone. We are going to have get the fuck out of here right now."

"What are you talking about? Raya stepped inside and looked at the open cabinet, then walked over to the crack pipe drawer to check on those items. "What are you talking about? Kaley probably got out of the cabinet and took off somewhere. Let's go look for her." Raya wanted to get high first.

"Are you that fucking stupid, girl?" Gallardo asked. "The front door was opened with a crowbar. Do you think Kaley picked the cabinet lock from inside the cabinet and then climbed out a window only to find a crowbar to smash the lock off our front door by herself? Are you brain-dead? Somebody broke into the trailer, found Kaley, and took her. I'm sure the police are on their way here or will be on their way here shortly. I don't know exactly what happened or who took Kaley, but staying here to figure it out is not very smart unless you want to end up in jail for child abuse." Gallardo spoke clearly and his intoxicated state did nothing to impair his assessment of their troubles and did nothing to impair the performance being recorded on the nanny-camera above the sink. The damaging images of

Gallardo and Raya, along with the incriminating audio from the pair were being downloaded in real time onto Deklan Novak's laptop at the Hotel Steamboat Springs. Carmen Gallardo suddenly looked like a ghost had overtaken his body. His boorish demeanor transformed instantaneously to fear.

"Where is the laptop?"

Behr and Deklan pulled into the hotel parking lot before 9:30 p.m. Behr and Kaley went up to the room. Deklan walked across the parking lot to a nearby mini mart/gas station. Deklan loaded up on premade sandwiches, cereal, milk, ice cream, and Little Debbie chocolate cupcakes, Swiss rolls, and strawberry shortcake rolls. The menu would not have won a blue ribbon at the National Alliance for Nutrition and Activity or the Institute of Child Nutrition, but Deklan thought the haul might look like heaven to Kaley. When Deklan got back to the room, Kaley looked confused but most of the fear in her eyes had dissipated. The little girl clutched her matted bear and sat quietly on the bed. Deklan brought some treats over to Kaley and she ate two chocolate cupcakes quickly before someone decided to take them away. Kaley knew her father, but Behr recognized she had to wonder why her father had allowed these things to happen in the trailer. Behr sat next to Kaley and slowly held her closer. Nothing was going to be forced. Behr sat there embarrassed and humiliated that he had allowed this to go on.

Deklan looked at the laptop. The nanny-camera connection was clear, but Gallardo had not returned to the trailer yet. Neither had Raya. The computer site was recording. Deklan knew that what they needed to record on the video feed was going to occur very soon.

The Steamboat Springs Police Department (SSPD) received the call at 11:30 p.m. Two officers and a detective arrived at the Hotel Steamboat Springs just before midnight. CPS had been notified and a representative was in route to the hotel. Middle of the night calls were not common, but the agency had procedures in place that were actionable 24 hours a day. Behr and Deklan had considered taking Kaley to the van in the parking lot. Deklan had no idea what law enforcement was going to do with the circumstances that were about to be presented concerning a child kept prisoner in a cage. How were the authorities going to assimilate an imprisoned child discovered by her father and a friend paying an unannounced visit at 9 p.m.? The two men decided that hiding the child from the police was a bad plan, but almost everything surrounding the entire plan was far from ideal.

Nothing about the story was untrue, but everything about the story was very hard to believe. Child abuse, child imprisonment, food deprivation, parental induced illness, fraudulent insurance policy, multiple trips to the emergency rooms for fabricated illnesses induced by the mother and boyfriend, not to mention the unlocked firearms and drug paraphernalia in a child's home. Deklan was prepared to present photographs and videos to back up the story. The police would question Behr and Kaley separately. Deklan knew what was coming. CPS had the right to speak to the child without a parent present but considering the unusual circumstances present, Behr and Deklan hoped the interview would happen while the father was with his daughter. CPS and the SSPD were about to determine if Kaley was going to be taken into child protective services and if Behr and Deklan were about to be charged with any number of potential crimes, not the least of which included breaking and entering as well as kidnapping.

Once CPS had arrived at the hotel, Deklan and Behr recounted the events of the evening and the events of the past year that led up to the most recent visit to Colorado Springs. Behr did most of the talking and referenced Attorney Dwayne Flynn at Northwestern University Law School as his legal advisor since the time of the divorce. The Steamboat Springs Police Department on site detective reminded Behr and Deklan that while they were not under arrest, they were allowed to have an attorney present during questioning. The Fifth Amendment states: "No person shall be compelled in any criminal case to be a witness against himself." The statute applies to those arrested and those in situations where they are being questioned by law enforcement but have not been arrested. Behr Thomas and Deklan Novak had decided that Kaley's safety and well-being was the only goal in mind.

The full recollection of the evening unfolded as well as the background story of Kaley's medical journey through the hospitals and medical centers in Steamboat Springs and the surrounding communities within a 2-to-3-hour drive from Steamboat Springs. The photographs and video evidence from the evening were presented to the police and CPS. Deklan handed over the laptop he took from the mobile home. Kaley was introduced to the authorities present. Behr had spent more than 30 minutes talking to Kaley alone, while the CPS case worker and the detective waited. Kaley was finally ready to speak to the people in the room.

CPS case workers knew the difference between general questions and the questions that would eventually be explored by professional interviews with child psychologists and legal representatives of anyone charged in the case. The questions directed at Kaley in the hotel room were not leading, but generic to the well-being of any child in an adult's care. CPS first needed to report on the physical appearance of the child. Since the Steamboat Springs detective on scene was a female, the detective was able to examine Kaley before the interview. Notes were made prior to the interview on Kaley's physical condition. The CPS case worker might request a further examination at the hospital if injuries or the child's condition warranted another exam.

Child Protective Services had to obtain a court order to call for an examination at a hospital or have the parent's consent. In most cases, CPS wanted the voluntary hospital examination to be conducted on the same day as the initial interview. Kaley's weight and gaunt, haggard appearance were sufficient visual clues to request a full examination. Injury patterns in suspected child abuse cases were generally pathognomonic or indicative for mistreatment or molestation. Since the suspected abuser was not present, the common explanations of accidental vs inflicted injuries were not debated. The condition of Kaley was ample justification to request an examination regardless of whether there were bruises or wound evidence present. Behr Thomas was asked before the interview began if he would authorize the examination. Behr agreed to a hospital examination

CHAPTER TWENTY-TWO

after a brief conference with Kaley and Deklan. Steamboat Springs Police Department Detective Ashley Grammer had been very careful with Kaley during a brief preliminary examination. Kaley sensed something safe when she spent a few minutes with Detective Grammer. Before the Child Protective Services interview got started, Detective Grammer had a couple questions for Kaley.

"The other lady here from Child Services, Miss Andrews, is going to ask you some questions like we talked to you and your father about a few minutes ago, Kaley." Detective Grammer explained. "Before she does, I need to ask about when you first got out of the locked cabinet earlier tonight. Is that okay, Kaley?"

"Okay." Kaley replied with her head down.

"When did you first hear or see your father and his friend, Deklan Novak?" Detective Grammer asked.

"W-What?" Kaley looked up confused.

"When did you first see or hear someone else at the trailer tonight?" Detective Grammer tried to clarify without leading Kaley anywhere specific. "Do you recall when you first knew that someone other than your mom and her boyfriend was at the trailer?"

"I'm not sure." Kaley stuttered a little and was very confused, afraid she might say something that would get her in trouble again. The questions were scary.

"That's okay, Kaley. Don't worry about that now." Detective Grammer reassured the frightened six-year-old. "Miss Andrews is going to ask you some questions and I promise, whatever you say will not get you in any trouble. Trust me, Kaley. I have a daughter your age and all of us are here to make sure nothing else bad happens to you." Detective Grammer leaned over and kissed Kaley on the forehead gently. The Child Protective Services case worker, Samantha Andrews took over for the Steamboat Springs PD Detective.

"Thank you so much for talking with me." The CPS case worker began. "Can you describe or detail what happens to you on a typical day with your mom and her boyfriend, Mr. Gallardo?"

"Okay." Kaley replied but wasn't sure what to say next. Kaley remained silent.

"What time do you normally get up in the morning?" The case worker helped Kaley understand. "Where do you sleep at night? Let's start there."

"I have a bedroom in the trailer home." Kaley said softly.

"Is that where you always sleep?"

"Yes…but." Kaley's voice trailed off.

"What do you eat in the morning, Kaley? Who makes your food?"

"Sometimes I have cereal in the morning. Mommy gets me the cereal." Kaley answered.

"Sometimes? Does that mean that you don't always eat breakfast?"

"Yes. Sometimes I don't eat in the morning."

"Have you ever gone all day without any food?"

"Yes." Kaley replied and did not explain further.

"Where do you play during the day, Kaley?"

"I play mostly in my cage." Kaley replied.

"You have a cage?" The case worker wanted clarity as much as possible.

"It's like a cage, but I have to stay there when everybody leaves the trailer." Kaley answered and looked at Behr. Behr nodded and wanted to be next to his daughter. The interview required Kaley to sit by herself…to eliminate the appearance of any coaching or rehearsed responses.

"Do you get to play with any friends from the trailer park or from school?" The case worker did not want to dwell on the cage topic. The cage topic would be front and center at a trial if a trial became necessary.

"I don't go to school." Kaley responded.

"Do you have friends from the neighborhood?"

Kaley looked confused and wasn't sure how to answer the question. The CPS case worker rephrased the question.

"Do you go outside every day and see other kids from the neighborhood?"

"No." Kaley did not elaborate. The case worker moved on.

"Does your mommy home-school you?"

Kaley looked confused and could not answer. Kaley had no idea what home-schooling meant. After a short silent pause, Kaley replied again. "I don't go to school."

"Does any place on your body hurt, Kaley?"

"Sometimes." Kaley answered.

"When does your body hurt, Kaley?"

"Mostly when I get sick, and mommy has to take me to the hospital."

"What happens at your house when mommy or mommy's boyfriend get angry?"

"Sometimes he grabs my arm really hard and sometimes I get hit." Kaley appeared to be more open when that question was asked. "Mommy gets hit, too" Kaley added.

"Where does mommy's boyfriend hit you?"

Behr Thomas felt his heart race and the anger surging throughout his bloodstream began to impair his breathing.

"Mommy gets hit in the stomach a lot." Kaley recalled dispassionately. "Daddy hits me there too."

"You call mommy's boyfriend daddy?" The case worker was mildly surprised.

"He makes me call him that." Kaley said.

"You know he is not your father, right?"

"I know." Kaley agreed.

"Who is your father?" The case worker asked. Kaley pointed to Behr Thomas. "That's right, Kaley."

"How often do you get hit, Kaley? How often does your mommy get hit?" The CPS case worker proceeded slowly.

"We get hit when mommy's boyfriend gets mad." Kaley replied.

"Are you afraid of anyone in the house, Kaley?"

"Yes." The answer was short, but direct.

"What happens when you take a bath?" the case worker asked.

Kaley didn't answer. Kaley looked down at the carpeting in the room. Behr stood up, but the detective motioned for him to sit down.

"What happens to you when you go to sleep, Kaley?"

Kaley continued to look down at the ground. The case worker had been schooled in preliminary investigations and was well-versed in where to stop and where to reasonably assume that a report and an investigation was necessary. The facts presented by Kaley and her father were sufficient to indicate that the child being interviewed was in imminent danger from her mother and the boyfriend living at the trailer. The case worker would recommend that Kaley be allowed to stay with the nearest living relative, in this case, her father was the closest relative. Behr's out-of-state residence was irrelevant to the case.

Everything about a question/interview session with a child had to be done correctly. Frequently and in this case, the child was the only one who could provide a reliable account of what happened at the trailer on a daily, weekly, and monthly basis. The legal system needed the minor to be a reliable witness and provide accurate and non-hesitant chronological accounts of the events. When discussing a delicate subject, a poorly worded question jeopardized the validity of the answer. In addition, a poorly executed interrogation often brought back the child's traumatic experiences so vividly that the child felt victimized again. The videotaped interview in each case was examined for body language, changes in the child's facial expressions, wincing at certain questions, and the physical reactions by the child to certain questions. The questions were scrutinized as much as the child's answers in a courtroom.

The CPS case worker and the Steamboat Springs Police Department had sufficient evidence to obtain arrest warrants for Raya Nolan Thomas and Carmen Gallardo. Kaley Thomas stayed with her father. Behr and Deklan brought Kaley back to Illinois two days after the interview with Kaley was conducted. Raya Nolan Thomas and Carmen Gallardo were arrested in Steamboat Springs and booked on Felony child abuse charges. According to the Colorado revised Statutes Section 18-6-401, a person commits child abuse if such person causes an injury to a child's life or health, or permits a child to be unreasonably placed in a situation that poses a threat of injury to the child's life or health, or engages in a continued pattern of conduct that results in malnourishment, lack of medical care, cruel punishment, mistreatment, or an accumulation of injuries that result in the

CHAPTER TWENTY-TWO

death of a child or serious bodily injury to a child. Penalties for felony child abuse in Colorado carry sentences that ranged from 2-16 years in prison. Neither Raya Nolan Thomas nor Carmen Gallardo were able to post bond. The pair remained in jail when Behr, Deklan and Kaley arrived back in Evanston, Illinois.

CHAPTER
TWENTY-THREE

NOVEMBER 2014...THE DAY BEFORE THE NOTRE DAME GAME

THE WILDCATS ARRIVED in South Bend on Friday afternoon. They went to the stadium for a walk-through practice and were bused back to the hotel. The team itinerary for a normal road game was as follows:

Position meetings (offensive, defensive, and special teams…also broken down by position – wide receivers, linebackers, O-line, D-line, DB's, backs, etc.) upon arrival at the hotel…5 p.m.

Dinner at 6 p.m.

Chapel at 7 p.m.

Team movie/pool stretch/players relax…7:30 p.m.

Team snack at 9 p.m.

Lights out by 10:30 p.m.

The only full team meeting occurred at the stadium on gameday prior to the game. Coach Adams decided to treat the trip to South Bend differently. The game was bigger than any other game during the year.

"We treat each and every game as the most important game of the year." Bullshit. The Notre Dame game is way bigger than the others." Adams told the team, breaking the coaching cliché dribble rule.

The recent Chicago Daily-Times article about Deklan Novak was another unusual focus to the weekend. Adams wanted the attention directed to Deklan addressed and dealt with before the game. When the players arrived back at the hotel following their brief walk-through at Notre Dame Stadium on Friday evening, Coach Adams called for a full team meeting to be held before the team split up into their position meetings. If the full team meeting lasted until dinner, then the position meetings would be held on Saturday morning before the team left the hotel.

Deklan Novak had not been asked about the article by his teammates or the coaches. There was an unspoken aura around Deklan Novak. The players felt it. The coaches would not admit it, but they were aware of the same sense regarding Deklan Novak. No one fucked with Deklan Novak and that was from respect for his service. That respect trumped everything normally associated with male sports teams. The naked rope climb was generally understood to have been a summer camp tradition that Deklan had allowed to happen. The naked "human carwash" and a punishment called "Shrek claps," were understood to be off-limits for the Navy SEAL linebacker... more out of self-preservation than military reverence. The usual hazing silliness that colonized every football locker room from Pop Warner to the NFL was unconsciously dialed down when Deklan was around.

Deklan filed into the banquet room at the hotel for the full team meeting once the buses had arrived. Everyone knew why the meeting was called and no one said a single word about the elephant in the room. When the dust settled and the thundering herd of players and coaches found their marks, Head Coach Lance Adams stood up to speak first.

Coach Adams began, "I don't have to tell anyone on this team about the importance of the game tomorrow night. We are here to win the game and the way we win the game is as one team. We are members of the Northwestern University Wildcat football team. Every man in this room is your brother and when someone attacks one of your brothers, they attack all of us. The Chicago Daily-Times

CHAPTER TWENTY-THREE

newspaper and a reporter named Xander Moss attacked one of your brothers this week. The paper asked for an interview from Deklan Novak. I asked Deklan if he was willing to sit down with the reporter. Deklan declined, not from a desire to hide anything but from the foundation of character that had been learned in the Navy and pushed to another level as a member of this football team. Moss and the Chicago Daily-Times published the article without Deklan's input or comments. I did not ask Deklan about any of the assertions or questions alluded to in the Daily-Times article. I asked Deklan if this story could hurt our football team in any way. Deklan told me that he would quit the team today if he thought for one second the story would hurt the team. Gentlemen, that was all I needed to hear from Deklan Novak. Our mantra regarding any questions from the media about Deklan Novak and his connection to the story from the Chicago Daily-Times is no comment. The Northwestern University Wildcat football team will address football related questions only. Deklan asked me if he could speak to the team briefly this afternoon. Deklan, come up here and say your peace."

Coach Adams walked away from the podium and stood waiting for Deklan to make his way up to the front of the room. The traveling team was comprised of 70 players, plus coaches, staff, and many front offices school personnel that were invited to the Notre Dame game as way of thanking many of the people who make a football program so large run flawlessly. The entire entourage had crowded into the hotel banquet room and had grown to nearly 120 people. Not certain of the protocol involved, when Deklan began walking to the podium, a few staff members began to applaud. A few more joined in and by the time Deklan stood before the room, everyone was standing on their feet, cheering like a high school homecoming game. The Northwestern University football family made Deklan Novak do something Navy SEALs rarely do. Novak blushed.

"Thank you so much for the unexpected vote of confidence," Deklan began. "I am not good as a public speaker so my remarks will be brief and anyone in this room can ask me anything they want at any time down the road. I am honored to be a part of this team. I had the honor to be a part of something incredibly special in the Navy. I feel every bit as humbled and grateful to be a part of something incredibly special at Northwestern University. Nothing I do will ever bring shame or scrutiny to this family. I cannot speak for others if they try to bring any of us down, I am here for each guy on our team. With respect to the article in the Chicago Daily-Times, I have never met or spoken to Xander Moss, the reporter who wrote the story. Three high school seniors disappeared after they graduated high school in Winnetka, Illinois in June 2000. They left the day after the graduation ceremony on a golf trip to Myrtle Beach, South Carolina but never arrived at their destination. No one has ever determined what happened to the boys. One of the boys on that trip was my stepbrother. His name was Andrew Stanley. My father had an extramarital affair in 1981, the year before I was born, and that relationship produced a child. I never knew about my stepbrother until we were both defensive starters in the same area during high school. I had read about the stellar Trevian D-lineman in the local papers. I found some letters in our house after my sophomore year in high school that detailed the relationship between my father, Andrew's mother, and the baby that my father helped support. My mother and my father never mentioned a word to me about the relationship or a sibling. When I found the letters, I saw no reason to tell my parents, but I did want to know my brother. We met and spoke on numerous occasions during our last two years of high school. Andrew went to New Trier High School in Winnetka, Illinois and I went to Loyola High School in Wilmette, Illinois, less than two miles apart. Our schools did not compete in the same conference, and we did not compete or meet in the state high school football playoffs. Andrew was a bad-ass D-lineman/linebacker and had a scholarship to play for Clemson University and head coach Tommy Bowden. An arrest prior to his senior year of high school ended Andrew's chance to play for Clemson University. The charges were eventually dismissed, and Andrew had made plans to play D-2 football for a year or two and then transfer back to D-1. Andrew never got the chance to follow that plan." Deklan paused and drank some water.

CHAPTER TWENTY-THREE

"Obviously, I have no clue what happened to those boys. I lost a brother I barely knew. Xander Moss never met Andrew Stanley and has never met me. Any theories or fairy tales surrounding the connection between me and the missing boys is fabricated. Open cases seem to attract as much speculation from a decorated reporter as the cases get from Hollywood. I apologize for bringing the unsolicited attention to our program, but I have no control over what is reported in the media. My circumstances have attracted more attention than I imagined or wanted. The only thing we should be concerned about is the Notre Dame Fighting Irish tomorrow night at 7:00 p.m. My personal and educated opinion is that Fighting Irish will be rolled into the next fucking county by halftime tomorrow night, but that is just my fucking educated opinion. Does anyone else in this room feel the same fucking way as I do?"

The room exploded with a thunderous ovation and the muscle-bound assembly began jumping out of the seats they occupied. Banquet hotel employees summoned the hotel security guards to quell whatever was happening in the banquet hall. Deklan Novak sought to become a small part of a D-1 football program. Circumstances and hard work opened the door and placed Deklan on the perfect team. Now, instead of becoming a small part of a football program, Deklan Novak had become the face of the Big Ten. Veterans from every branch of the military and their families, no matter what Big Ten team their allegiance rested with, suddenly became Wildcat fans for the longevity of Deklan Novak's eligibility. Xander Moss and his article had not started a criminal investigation or a renewed law enforcement discussion to uncover the conditions relevant to the 2000 cold case he was obsessed by. Xander Moss launched a quiet, reluctant Navy veteran into the role he never sought but ardently accepted.

CHAPTER TWENTY-FOUR

2014...THE TRIAL IN STEAMBOAT SPRINGS

FROM THE SPRING of 2014 through the fall, the case against Carmen Gallardo and Raya Nolan Thomas in Steamboat Springs, Colorado fell apart. Once the evidence discovered at the trailer home located at the Dream Island Mobile Home Park was tainted and deemed inadmissible in court, the efforts to salvage the case were in vain. Following Kaley's admission to the Steamboat Spring's detective, Ashley Grammer, that she could not recall how she came to see Deklan Novak for the first time, the Novak testimony about entering the trailer home because he heard the cries of a child in danger, was thrown out. The discovery of the cage, the filthy conditions inside the cabinet/cage, the small cleaning bucket used for a toilet, the total darkness, the countless hours of leaving a six-year-old child alone, the drug paraphernalia, the unlocked firearms and ammunition in the home, the laptop computer, and the Ipecac bottles found in the bathroom were all discovered because of a break-in or an illegal entry. Every piece of evidence discovered after Deklan entered the mobile home was inadmissible evidence. The only relevant evidence that remained actionable was Kaley's physical condition, which was reported on by Child Protective Services case worker Samantha Andrews.

Carmen Gallardo was released, and all charges were dropped. Raya Nolan Thomas had her charges reduced to a Class 2 misdemeanor in the Colorado Criminal Code (CRS-18-1 .3-501). Released from jail on her own recognizance, Raya faced a maximum penalty of 120 days in jail (already served) and/or a fine of up to $750.

The fraud committed to forge an insurance policy that paid Raya in the event of Kaley's death, which had been an ongoing course of action carried out by Gallardo and Raya, was tainted as well. The policy was signed over emails between the insurance agent and Carmen Gallardo, posing as Behr Thomas. Since Behr Thomas was a good customer and his account was in good standing, adding a policy to the existing client account was made easier by using Docusign, an internet organization that allowed individuals and organizations to manage electronic agreements with electronic signatures on different devices. Signatures processed by Docusign were compliant with the United States ESIGN Act (2001 Electronic Signatures in Global and National Commerce Act...Federal Trade Commission). Carmen Gallardo was free and clear of all charges. Raya Nolan Thomas faced a misdemeanor with no jail time possible because she served enough time when she was originally charged and could not post bail. No charges were filed regarding the bogus insurance policy that was set to reward the couple after they killed Kaley, slowly and methodically. The only positive aspect of the case at the time was that the judge did not return custody of Kaley to Raya. The court instructed Raya to file a challenge to the order from Child Protective Services that removed Kaley from the home.

A dispositional order and the removal of Kaley from Raya's home subsequently was challenged by a pro-bono legal team based in Boulder, Colorado. Colorado Legal Services provided access to high-quality legal services in the pursuit of justice for as many low-income people throughout Colorado as possible. They were able to assist with family law, housing issues for seniors, difficulties with government programs, tax problems with the IRS and other civil problems.

CHAPTER TWENTY-FOUR

Colorado Legal Services (CLS) took the case for Raya Nolan Thomas. The brief was filed before Judge Judy Danilson at the Routt Combined Court, 1955 Shield Drive Unit 200 in Steamboat Springs, Colorado. The petition contended there was not clear and convincing evidence supporting the grounds for adjudication of the child as a child in need of assistance (CINA). The petition further claimed there was insufficient evidence to order the removal of the child and reasonable efforts to achieve reunification were not provided. Because there was no evidence to substantiate the order to remove the child from the home and the mother had been active in the care and sustenance of the child, and danger of harm to the child had not been established therefore any danger was not imminent, the order to remove the child from the home should be rescinded and the child should be allowed to return to her mother.

The brief went on to say that since no evidence of factitious disorder imposed on another had been established, as mentioned in the order to remove the child, the mother's mental health was not at issue and there was no evidence to conclude that factitious disorder imposed on another would in any way prevent the mother from the employment she had at the time of the order to remove the child from her home. There was no evidence that the mother had been unable to pay the rent and utilities, no evidence that the residence was in disrepair and no evidence that the mother kept an unreasonable number of animals in her home.

The State of Colorado and Child Protective Services put up a strong defense of the removal order based solely on the child's condition. CPS entered evidence that Kaley had been to numerous emergency rooms in the county over a short period of time, the time frame after Raya and Kaley arrived in Steamboat Springs following the divorce. CLS contended that Raya was a good mother and did not hesitate to seek medical help when her daughter became ill. CLS contended, thank God, Raya was an attentive and caring mother.

CPS did provide the court with home-school violations that included the lack of any kind of filed curriculum, any evidence that the child had been tested as required and no evidence of a qualified adult in the home that was teaching the child the state's mandatory minimum requirements to pass standardized tests. A six-year-old child was supposed to be in the first grade. There was no evidence that Kaley had any schooling. Kaley told the CPS case worker that she had not been given any books or school supplies of any kind.

The review of the case by Judge Judy Danilson was a textbook case of what the law allowed versus what was the obvious clear and best interest of the child. The ruling by Judge Danilson read:

"We are not bound by the juvenile courts, and we are not bound by Child Protective Services, however, we do give them weight. Our primary concern is the child's best interests. CINA (Child in Need of Assistance) determinations must be based upon clear and convincing evidence. In this case, clear and convincing evidence does not exist. Pursuant to Colorado Code section 232.5 (5)(©)(2), a CINA means a child who has suffered or is imminently likely to suffer harmful effects as a result of a failure by the child's parents or parent. The court does acknowledge the lack of clear and convincing evidence, but there has been a significant history of lying to medical professionals regarding the past medical history of the child. On numerous occasions, the medical care professionals at the county's emergency rooms were not informed of the child's previous visits to local hospitals. In addition, the child's physical condition at the time of the CPS interview was gravely concerning. Obviously, the child had not been getting enough food, had not been provided adequate care and provisions. I am aware of the evidence that was inadmissible, but the child also admitted to the lack of food and clothing in the interview. The court will leave custody of the child in the care of her father for 12 months. If the mother would like to file another petition for custody in 6 months, the court will hear the petition, but if none of the concerns brought up today can be eliminated then the petition will be denied. As I said earlier, the court's main goal is the best interest of the child. The only skin in the game we have is the child's welfare. I am not going to place the child back in the home where she couldn't get a decent meal, where the amenities within the home were not conducive to raising a six-year-old girl and to where the education of the child had been compromised completely. While the mother's behavior in certain areas did not present life-threatening behavior, the court has concerns about other occupants of the home. Further examination of the case only leads the court to work to ensure that no further harm comes to the child. We, therefore, deny the request to vacate the custody order imposed by Child Protective Services. The child will remain with Behr Thomas, the father of Kaley Thomas." The judge had another finding to read.

"The court further finds that Child Protective Services has made reasonable efforts towards reunification. The child's mother was incarcerated for months following the custody order to remove the child and the child now lives in Illinois. The petition's claim that the state did not make weekly visits to the mother, did not provide psychiatric assistance, and conduct regular evaluations were all due to impending felony trial requirements and the fact that the mother was not available. We conclude clear and convincing evidence supports the court's adjudication of Kaley, as a child better off with her father at the present time and we affirm the order from Child Protective Services to award custody of Kaley Thomas to her father, Behr Thomas and the child will reside in Evanston, Illinois." AFFIRMED

Northwestern University had a bye week following the Notre Dame game in 2014. Coach Adams gave the players a substantial break during the break-week to recover physically from the weekly grind of constant contact and to reenergize the focus that was often yanked in many directions for an 18–21-year-old student athlete. Northwestern was a tough school academically. Big Ten football was a full-time commitment on top of the academic requirements to stay on the team and to stay at the university. The time off was welcome for most of the student athletes. The 34-year-old Navy SEAL was in the weight room on the first Monday of the break-week. Deklan was throwing up 225 lb. bench presses in sets of 20. Behr Thomas came into the weight room and sat next to Deklan Novak.

"The case against Gallardo and Raya is toast." Behr started the conversation. Deklan put down the barbell and sat up. Behr continued. "All charges against Gallardo were dropped. The charges against

Raya were dropped to a Class 2 misdemeanor. Gallardo is a free man. Raya is free and out of jail. She doesn't face any jail time because if she is convicted of the misdemeanor, the maximum penalty is 120 days in jail, and she already served more than that because they couldn't post bail. The charges related to the insurance fraud were also dropped. Since the policy was signed by Docusign over emails, the forgery cannot be established enough to bring a felony indictment from a Colorado Grand Jury. The District Attorney in Cook County told me that there is no case worth pursuing with internet signatures as the case foundation."

"Let me be clear, Coach." Deklan followed up. "Carmen Gallardo is a free man with no charges pending?"

"Correct." Behr replied and waited for more.

"Raya, your ex-wife, who had been systematically trying to kill your daughter is now free and will spend no more time behind bars?"

"Correct."

"What about all the evidence that I found in the trailer?" Deklan was dumbstruck. "Gallardo and Raya were holding Kaley as a prisoner and doping her! There were unlocked firearms in the trailer. There was drug paraphernalia in the open. If someone was making a training video for Case workers at Child Protective Services, they could have used Raya's trailer as the video's message on what constitutes child abuse. What happened to all the evidence?"

"All the physical evidence and all the evidence that you collected with your camera was tossed because Kaley could not establish the fact that she called out for help and that was the reason you had to break into the trailer. The breach of the front door was deemed illegal, therefore, all the evidence obtained after you entered Raya's trailer is inadmissible in court."

"Unfuckingbelieveable. Are they coming for Kaley?" Deklan asked the next logical question stemming from the illogical set of circumstances presented.

"The Steamboat Springs Police Department called in the FBI to look at the laptop because they believed there was a good chance the laptop had been used in the insurance fraud portion of the case or

CHAPTER TWENTY-FOUR

may have been involved with child abuse. Kaley was not kidnapped from Illinois, but her confinement constituted false imprisonment, and the federal implications called for the FBI to look at the case. The nearest FBI field office to Steamboat Springs was in Glenwood Springs, about 114 miles and a two-hour drive from Steamboat Springs. Two Special Agents from the FBI arrived in Steamboat Springs and reviewed the evidence that you brought to the SSPD. The FBI took the laptop and sent it to the FBI Crime Lab in Quantico, Virginia. The analysis took a couple of weeks. Carmen Gallardo had downloaded naked photos of Kaley and was selling them on a website called Playpen on the darknet." Behr Thomas could barely finish what he had rehearsed telling Deklan.

"My God." Deklan blurted out. "Please tell me that the laptop evidence is admissible in court."

"None of the evidence is admissible in court, including the child pornography. We took the laptop from the mobile home after you "illegally" entered the premises. No photographs taken in the home, and nothing taken from or found in the mobile home is admissible evidence." Behr stared at Novak and allowed the words to find a resting place.

"They both walk?" Deklan was talking to himself.

"Raya tried to reverse the custody order. A pro-bono legal service in Colorado represented Raya and filed a formal petition to reverse the custody order, but the Judge in Routt County upheld the custody order issued by Colorado Child Protective Services. Kaley is staying with me for the next year at least unless Raya finds a way to get another judge to vacate the order." Behr explained.

"How likely is that to happen?" Deklan asked.

"Unlikely from what I have been told by the case worker at CPS. But this assurance comes from the state where they just let two people who tried to kill a six-year-old child, off scot-free. Raya was sent to a state-funded, non-profit rehab center that the pro-bono legal service in Colorado helped Raya locate. The center is in Silverton, Colorado, not far from Steamboat Springs. Who knows what Raya will be able to do if she cleans up? Am I relieved and confident that Kaley is out of danger from these animals? No, I am not confident at all that my daughter is safe from her mother and Gallardo." Behr was adamant.

"Do you think they will come here and try to take Kaley?" Deklan asked a rhetorical question.

"No clue, Deklan." Behr was tired from the season, tired of his own failures as a single dad, and tired of dodging the legal system's roadblocks to common sense.

"Gallardo isn't stupid enough to come here." Deklan offered. "He's a shitbag, but I can't imagine he would come for Kaley on your home turf."

"Maybe not." Behr replied. "But I never imagined my ex-wife could plot to kill our daughter, concoct a semi-coherent plan with Gallardo, systematically enact the plan in order to benefit from an insurance policy, and literally beat the legal system while she's addicted to crack and doesn't have a working brain cell left in her fucking head."

"Kaley was a meal ticket for Gallardo." Deklan added. "He won't go out of his way to get Kaley. She is no asset to him any longer. Bringing Kaley back to Colorado is another mouth to feed and the authorities will be all over his ass anyway. Don't worry about Gallardo. He doesn't have the balls to come here."

"You're probably right." Behr agreed. "In the back of my mind, I want him to come for her. A couple weeks ago, I got a package delivered to my condo in Evanston. The package contained a matted stuffed animal. I'm pretty sure that it was the stuffed animal that Kaley was clutching when you found her. I don't remember her having that animal after the CPS case worker conducted the interview that night. Maybe the bear got tossed? Maybe there was more than one stuffed animal at the mobile home? Maybe I just received a random matted stuffed animal by happenstance?"

"Maybe." Deklan responded sarcastically. "What can you do?" Deklan asked.

"Nothing." Behr conceded. "I can do nothing if I want to keep Kaley. The case worker at CPS told me to keep a low profile, make sure Kaley is in school, fed well, gains weight and stays out of any hospitals. The Judge went out on a limb and affirmed the order to

CHAPTER TWENTY-FOUR

move Kaley back to Evanston when the evidence told her to legally rescind the order. I didn't mention the FedEx package to anyone."

"How long have you known about the case falling apart?" Deklan let the FedEx package discussion slide.

"It has been ongoing for the last month." Behr replied. "I waited until the case was finalized before I came to you. You have been a good friend during this nightmare, and I hesitated to bring you the bad news, but once the Judge allowed me to keep Kaley, I wanted to bring you in on what has transpired. Sucks, huh?"

"Sucks big time, Coach." Deklan agreed. "Can I do anything?"

"Na." Behr stood up and patted Deklan on the back. "Keep up the 225's if you want to cement a place in the starting lineup. You are here working out when the rest of the defense is at The Keg getting hammered. Stick to the plan is what my college coach always told me. Stick to the plan, PO1 Novak." Behr turned to leave the weight room.

"Hooyah, Coach." Novak grabbed the barbell and pushed it towards the ceiling.

CHAPTER
TWENTY-FIVE

2014...THE BYE WEEK FOR NORTHWESTERN UNIVERSITY FOOTBALL AND A BARROOM BRAWL

DEKLAN HAD ARRIVED at Sunpies shortly after 7 p.m. It was cold in mid-November in Steamboat Springs. The patio that overlooked the Yampa River was empty even though the portable heaters were there to help increase the seating capacity year-round for the Steamboat Springs tavern. Sunpies had been the favorite watering hole for the locals since their opening in 2005. The owners came from NOLA and named the joint after a "famous" jazz singer, Sunpie. Sunpie and his band, The Louisiana Sunspots were regulars amidst the New Orleans jazz scene and the Bourbon Street madness that ran nonstop. Most tourists who ventured into Sunpie's imagined that the local hangout had been a decades long tradition in the ski town.

The oldest bars in Steamboat Springs were the Steamboat Cantina and Tortilla Factory and a couple mountain lodge bars that opened in 1963 when the Steamboat Ski Resort first opened. The tavern history in Steamboat Springs does not harken back to the days of Butch Cassidy and the Sundance Kid. Deklan thought Sunpie's was lacking in the old west vibe and had been expecting a few local taverns with exquisite, hand-carved back bars like the J-Bar at the Hotel Jerome in

Aspen or the trophy laden walls inside the Buckhorn Exchange in Denver. Where were the bars that dated back to the days of the Wild Bunch, a collection of outlaws who roamed the Wyoming, Colorado, and the Utah territories in the 1880's and '90's. Apparently Butch, Sundance, Kid Curry, Ben Kilpatrick, and Will Carver never hung out at Sunpie's. Deklan Novak was briefly distracted by the historical cavity that the soccer décor failed to fill but returned to the task at hand when Carmen Gallardo entered the bar.

Deklan wore a Denver Broncos hat that he picked up at the airport. Deklan wore sunglasses and a North Face winter parka. There was a short, thick beard and the only visible facial features on Deklan Novak were his nose and his cheeks. Deklan had never met Carmen Gallardo, so the semi-disguise was for other reasons. Since Deklan and Behr had first traveled to Steamboat Springs, Deklan remained reluctant to check a firearm at the airline ticket counter. Traveling with a firearm legally meant alerting the ticket agent of the weapon inside the checked bag before boarding. Passengers were allowed to travel with a firearm if the gun was unloaded, packed in a hard-sided locked case, and packed separately from any ammunition. The troubling aspect to the gun issue arose by creating a record of traveling with a firearm. The record could always be used to support an argument of premeditation. Deklan hated to call Pat Gabriel again, but he did. Pat Gabriel was going to meet Deklan Novak at the Hotel Steamboat Springs, and the two friends were going to ski for a couple days because Deklan had a bye week from the Big Ten schedule. Pat Gabriel also carried with him an extra Glock 17 with three fully loaded magazines. Pat and Deklan met in the late afternoon after Deklan arrived. They had a couple beers at Smoke & Rye, the bar located inside the Hotel Steamboat Springs. Pat Gabriel met a lady for dinner. Deklan went back to his room on the third floor. Deklan left the gun at the hotel when he ventured out alone on that first night in Steamboat Springs.

Gallardo was well-known at the bar. Bartenders and patrons knew him. Gallardo was a big guy in the Benecio Del Toro mold, a Mexican immigrant but could have passed for Puerta Rican. Both nationalities

CHAPTER TWENTY-FIVE

shared roots from Europe, Africa, and other ancestries. Gallardo resembled from afar the character Javier Rodriguez, minus the charismatic smile and the villainous scowl that was as treacherous as it was intoxicating, played by Del Toro in the 2000 film, **Traffic**. Deklan was nursing a second New Belgium Fat Tire, a delicious amber ale from a brewery in Fort Collins, Colorado.

Carmen Gallardo sat drinking a cheap whiskey and Coke at the bar. There were five other men near Gallardo, who appeared to be friends or barroom buddies with the ex-con former boyfriend of Raya Nolan Thomas. The men were blue collar drug mules and out-of-work laborers. They all dressed alike, as if they were auditioning for a part on American Loggers or Ax Men. Gallardo ordered another whiskey and Coke from the bartender, a husky dude named Colton James from Vancouver Island in British Columbia. Colton James loved to entertain the Steamboat tourists with wilderness tales from the north country. Gallardo and Colton James shared a laugh. Deklan got up and walked over to the bar next to Carmen Gallardo. Deklan tapped Gallardo on the shoulder. Gallardo turned to face Deklan Novak.

"What?" Gallardo snapped like the tough guy he envisioned himself to be.

"Did you see a ten-dollar bill laying on the bar here? I'm pretty sure I accidently left a ten-dollar bill here on the bar." Deklan asked and explained.

"I didn't see any money on the bar, pal." Gallardo replied rolling his eyes.

"I'm positive I left a ten-dollar bill next you on the bar by accident." Deklan repeated. "You didn't take it did you?"

"I just told you asshole that I didn't see a ten-dollar bill on the bar and if I did, I would have given it to the bartender." Gallardo chuckled as he looked back at Colton James. "Now, go back to your fucking little table before you get hurt."

"What's your name?" Deklan asked the guy sitting next to Gallardo. The guy was large and got even bigger when he stood up. The guy next to Gallardo sported a ten-inch mass of scraggly beard that hadn't been groomed since the growing process began. The 6' 2", 290-pound sasquatch was wearing a plaid red and black flannel shirt, a red down vest and a Navy SEAL ball cap.

"My name is fuck you." The man barked back, now standing.

"Were you a Navy SEAL?" Deklan asked. "Bad boys, those guys."

"Yeah, I was a Navy SEAL dipshit, but we don't talk about it." The evil stare grew more menacing. The Navy SEAL imposter's name was Chris Haley, a Texas transplant from Amarillo.

"But you wear a hat announcing that you were a Navy SEAL?" Deklan wondered out loud. "I get it. SEALs don't blow their own horns. What number BUD/S class were you in?" Deklan inquired.

"What?" Haley snapped.

"What number was your BUD/S class? Every BUD/S class has a number. What was your class number?" Deklan waited.

"Fuck you. None of your fucking business."

"Sorry pal." Deklan apologized. "I had some buddies about the same age as you who were SEALs. Maybe you knew one of them?"

"Take off dipshit. These boys here are a league you don't want to fuck with." Chris Haley suggested in that exaggerated Texas drawl.

"Cool." Deklan said and turned back to Gallardo. "Why don't you give me twenty bucks and we can call it even."

"You told me that you left ten bucks on the bar?" Gallardo's comment was a question as he looked confused momentarily.

"I did leave ten bucks on the bar, but it's going to cost you another ten bucks to avoid the smackdown you are about receive for taking my original ten bucks." Deklan smiled and scanned the room.

"Mother fucker!" Gallardo snapped and pushed his stool back while standing up. Gallardo pointed to the bartender. "Don't call the cops when this guy is on the floor. We'll leave him down by the arena and the river. They can drag him up from there and you'll keep the law out of the bar. Gallardo turned back to Deklan Novak. They were standing face to face now.

"You are alone here, asshole." Gallardo pointed out the obvious as Gallardo's friends pushed away from their barstools as well. "Chris, this is going to be fun." Gallardo patted Chris Haley on the shoulder.

CHAPTER TWENTY-FIVE

"Look around dickhead." Another friendly face chimed in. "There are 5 or 6 guys here that will take you apart fuck-brain."

Gallardo went first. The first punch that Gallardo reared back to throw was telegraphed eloquently. Deklan fragmented the attack in a split-second by charging into Gallardo's incoming right fist and then by raising a simple forearm block thwarting the initial attack. In a fluid and a lightning-fast counterattack, Deklan began by striking Gallardo at the base of his nose with a palm heel strike. Gallardo's nose exploded like a water balloon. Deklan then pulled his right hand back and locked his right arm underneath Gallardo's left shoulder while at the same time, locking his left hand behind Gallardo's neck. The MMA move gave Deklan complete control of Carmen Gallardo's body movement. Well-trained bouncers often used BJJ techniques to remove troublemakers from an establishment. Deklan moved Gallardo's body out away from the bar and then slammed Gallardo's head into the face of the bar by pulling down hard with his left hand and forearm. Gallardo's forehead slammed into the intersection of the bar top and the front façade of the bar. Blood poured from a gash over Gallardo's right eye. Novak repeated the collision, and his opponent fell to the floor in a heap, like a wet sandbag dropped from a second-floor balcony. Deklan felt the rear naked choke hold as one of Gallardo's friends came up from behind to help. With blinding speed, Deklan reached behind his head and grabbed the hand that was pushing Deklan's head into the lock arm of the popular Brazilian Jiu Jitsu move. Deklan pulled the hand off his neck, then pulled the hand over his head and into view. Deklan snapped the man's wrist down sharply while sending his opposite elbow back up with the force of a car-compactor, snapping the attacker's elbow. A primeval scream ripped through the tavern like an air raid siren. A dozen or so bar patrons froze after the scream. They had been moving to exit Sunpie's. The general American population is not a combative society and most often reacted with panic when the sounds surrounding them included a man's fist striking another man's face. The confrontational noise was quite distinct and frightening to the vacationing Caucasian couples in the tavern. The choice of a local hangout was bound to be reviewed later.

"Don't fucking move!" Deklan pointed to and yelled at the next man up, Chris Haley in the Navy SEAL hat considered his options. "If you take one more step closer to me, I will break your jaw in eight places, and you will be sipping your meals through a fucking straw until next Halloween." Deklan explained rather emphatically. Deklan Novak's veins were protruding from his neck. The broken elbow scream echoed in Haley's head. Chris Haley put up his hands in surrender mode, indicating he wanted nothing to do with the real Navy SEAL in the room. Gallardo's buddies backed away and reevaluated how close the friendship had truly been to the slumped figure on the floor.

Deklan gazed around the bar again. The bartender had not reached for the phone. Deklan pushed Gallardo back against the bar and out of the aisle. Deklan patted down Gallardo's jacket and pants. Novak headed for the door but stopped and addressed the stunned posse, a final time.

"Take his advice and don't call the police." Deklan started to leave Sunpie's, but stopped in the doorway and looked up at the outdoor surveillance camera covering the front door. There were two cameras inside the bar. One camera was an observation tool for watching the bartenders. The second camera inside the tavern covered the main floor area where patrons drank and socialized. Deklan pointed back to Chris Haley.

"Take that fucking hat off." Deklan ordered. The man complied. "Don't fucking wear it again."

Deklan didn't stick around for an affirmation. Deklan accurately assumed the idiots in the bar would take at least 2-3 minutes to search Gallardo and another few minutes to decide what to do. By that time, Deklan was long gone.

Gallardo struggled to regain his equilibrium. The ex-con pushed his friends away as they helped him up. Chris Haley stuffed his hat into a vest pocket while scanning the bar to see if anyone noticed. The man with the shattered elbow was moaning and staring at the

CHAPTER TWENTY-FIVE

grotesque angle that his elbow now held. Colton James called 911 for the shattered elbow guy, who was about to pass out. Gallardo looked around the floor area where he fell. Blood stained the floor. The suddenly furious ex-convict had been carrying drugs and wanted to see if anything fell out of his pockets. Gallardo patted himself down to locate his revolver and his stash. The .38 and the drugs were not missing. Nothing was on the floor except a plastic key card in a cardboard sleeve. The cardboard sleeve had the Hotel Steamboat Springs logo and a room number on the side.

"Definitely, do not call the police." Gallardo repeated his desire from earlier. "I will visit our guest later tonight by myself." Gallardo explained as he eyed the hotel key card that Novak had accidentally dropped on the bar floor during the fight.

CHAPTER TWENTY-SIX

LATE FALL 2014... THE SPECIAL OPERATIONS FORCES (SOF) TECHNIQUE CALLED THE MOZAMBIQUE DRILL

ROOM 303 WAS on the third floor at the Hotel Steamboat Springs. The room was directly across from the guest elevator bay. There were surveillance cameras on each floor that covered the elevator bays and the hallways. Deklan had left Sunpie's at 7:20 p.m. By the time Deklan made sure that no one followed him, he had circled the downtown area, dissected the streets at the Steamboat Ski Resort on Mount Werner near the Sheraton Steamboat Resort and then finally returned to his room. It was 8:30 p.m. Deklan spoke to Pat Gabriel briefly on his burner phone. Gabriel was going to be back in his room by 9:00 p.m. Deklan assumed that Gallardo was going to wait until later that night to pay him a visit. Deklan wasn't sure what time Gallardo was going to show up, but he was convinced that a visitor was in his near future.

The nightlife in Steamboat Springs was not South Beach, but there was ample activity throughout the downtown district of the popular ski town as the clock approached 11:00 p.m. Deklan thought Gallardo might set up surveillance in the parking lot of the hotel to pinpoint when his target arrived back at the hotel. Since there was no way to definitively know if Gallardo was observing the main entrance, Deklan left the hotel at 11:15 p.m. from the rear delivery dock and made his way down the block, returning on the main sidewalk, clearly visible from the parking lot. If Gallardo was watching the lot for Deklan's return, the seemingly late-night return to the hotel might trigger Gallardo's revenge outing. Again, Deklan had no way to know for sure, but after re-entering the hotel from the main entrance, Deklan peered out through a partially blocked window that was behind the concierge's desk. Lo and behold, Carmen Gallardo stepped out of a vehicle parked in one of the back rows of the parking lot, away from the overhead lights. Gallardo stood next to his car in the darkness and lit a cigarette. Fitting, Deklan thought, considering the hotel and its nicotine ambiance. Deklan returned to his room promptly.

 Deklan waited across from the small bathroom in a very small alcove next to and behind the main door to the hotel room. Once the door was opened, Deklan would be completely blocked from view by the open door. Deklan left a light on in the bathroom and the bathroom door was closed but not all the way. There was a small opening between the bathroom door and the door frame that appeared as if someone was in the bathroom. Deklan turned the water on from the bathroom sink and let it run. Deklan left the television set on with the volume audible but low, and a rerun of the Jimmy Kimmel Show was airing in Room 303. Just after 11;30, Deklan heard the elevator open on his floor. Deklan was next to the front door and could hear the elevator chime as it stopped and opened on the third floor. Deklan also watched the crease between the hotel room door and the floor. The lights in Room 303 were off, but the hallway lights cast a moving glow from the bottom of the door. Deklan remembered a scene from the 2007 film, **No Country For Old Men**, where Josh Brolin waited in his dark hotel room and watched the silhouette of Javier Bardem's feet stop in front of his room. Carmen Gallardo stopped in front of Deklan's room and stood silent as he listened to the television and made certain that Deklan Novak or the asshole from the bar, as Gallardo knew him, was in the room. The asshole was in the room. Gallardo palmed the revolver in his jacket and took out the key card.

Key cards made little noise as they released the door lock. The television was louder. Gallardo placed the key card back in his pocket and grabbed his revolver. Gallardo pushed the door open about two inches and stood silent. The television was on. Gallardo noticed the bathroom door was semi-closed and assumed the asshole was in the bathroom. Should he wait until the guy came out of the bathroom or surprise him? Gallardo burst into the hotel room space and rushed into the bathroom, gun drawn and yelling..." alright asshole, don't fucking move!"

No one was in the bathroom. Gallardo froze and was suddenly gasping for air. Gallardo's balance was shattered by the rear-naked choke and the gun that he was holding became immovable. Deklan had practiced the movements necessary to disarm an enemy hundreds of times in training as a student of Krav Maga and BJJ, in SEAL training and then countless times as an instructor at RTC Great Lakes. Deklan had executed the lightning fast Krav Maga and BJJ disarming techniques in combat at least a dozen times. While there had been more than two years since his last combat deployment, the muscle memory had not been lost. Speed and knowledge were devastatingly effective. As Gallardo tumbled backwards, Deklan released the choke hold and grabbed the barrel of the gun. Deklan had one hand on Gallardo's hand/wrist that held the gun and the other hand on the revolver barrel. The short violent snap of the gun broke Gallardo's finger in an instant because he had his finger on the trigger. The gun was immediately transferred into Deklan's possession as Gallardo gasped at the sudden pain. Gallardo fell to the ground, disorientated, disarmed, and stared at his revolver being pointed back into his face. Deklan Novak wore a pair of thin Tactical shooting gloves.

"Get up slowly, shit-bag." Deklan instructed decisively if not affectionately. "We are moving into the main room, and you are going to sit at the table by the window. Sit at the first chair you come to and for your own safety, do not make any sudden moves or entertain any thoughts of a counter-defense. I have done this for a living for years. You never stood a chance with me." The two men moved into the main area of the hotel room. Gallardo located the table and sat down.

"What are you, a cop?"

"You wish." Deklan replied but did not elaborate.

"Who are you?" Gallardo asked.

"I am Kaley's uncle." Deklan lied.

"Kaley didn't have an uncle." Gallardo stalled and stuttered. "How do you know Kaley? Who is Kaley?" Gallardo backtracked when he realized what was happening. "Who the fuck is Kaley?"

"Kaley is the girl you and Raya tried to kill." Deklan answered calmly. "Kaley is the six-year-old girl you locked inside a closet for hours and days and months at a time. Kaley is the little girl who you beat up and slapped around like you were Mike Tyson. Did that make you feel like a bad ass? What else did you do to Kaley that no one ever found out about?" Deklan stopped and placed the .38 revolver in the middle of the round table they were sitting at. Deklan took a chair opposite Gallardo. The gun sat on the table and Gallardo knew Novak was baiting him.

"I didn't do anything to Kaley." Gallardo announced.

"I thought you didn't know Kaley."

"What is this about?" Gallardo grew defiant. "All the charges against Raya and me were dropped. We didn't do anything wrong. I don't know what your connection to Kaley is, but I've got a crew of dudes ready to come up here if I don't show up back downstairs in the next two minutes."

"Cool. Let's wait. I can get my laptop, and we can check out Playpen. Maybe we can see photos of Kaley there?" Deklan said nothing more. Gallardo looked around the room and began to panic.

"What do you want?" Gallardo asked with sudden, windpipe constricting fear. Gallardo knew he was the wildebeest and Deklan was the lion, about to cinch his jaws into Gallardo's throat. Gallardo scanned his surroundings hopelessly for a clue to his next move.

"I want you to pay for what you did to Kaley." Deklan did not elaborate.

CHAPTER TWENTY-SIX

"What the fuck, dude?" Gallardo was running out of options. He stared at the gun on the table. Novak interrupted his escape plan.

"I'm going to tell you something that I haven't told another person. I never told my wife. I never told my friends, my mom, my coaches, my teammates, my platoon mates, or my commanding officers."

"Commanding officers?" Gallardo was confused. "What the fuck…"

"I am a Navy SEAL." Deklan mentioned calmly.

Gallardo's heart rate surged, his breath shortened, and his eyes darkened with some degree of immobility. Gallardo's options instantly became clear, yet the surging anxiety paralysis overwhelmed any inclination towards self-preservation.

"That's ten years asshole, as a member of a special operations unit trained to engage in direct action raids or assaults against enemy targets and terrorist groups." Deklan continued. "I served in combat zones all over the middle east. Afghanistan. Iraq. I lost track of how many combat missions I took part in over the years."

"What has this got to do with me?" Gallardo repeated and continued to stare at the revolver on the table.

"This has everything to do with you. As I said, I'm going to tell you something that I never told anyone else." Deklan stopped and thought again about what he was about to say. The conversation was years coming but Deklan had not planned to unveil his deepest, psychosomatic, therapeutic revelations to a child-abuser. Deklan had told his NU teammates that he hadn't killed anyone, but he knew they didn't believe him. "When I signed up to become a Navy SEAL, I knew my chances were near zero to survive the training. I didn't care. I wanted one thing to emerge from the toughest training in the world. I set out to become the baddest motherfucker on the planet in a good-guy uniform. I wanted to change the god damn world by killing bad guys. Do you know what happened when I finally got my chance to kill the bad guys?" Deklan paused and Gallardo didn't answer. Deklan stayed silent.

"No." Gallardo finally answered and sensed the cadence had turned ominous.

"I never killed anyone in combat." Deklan admitted to himself as much as anyone else. "I trained to kill terrorists, and I spent almost three years killing myself to become a trained assassin, with assignments to kill the worst people in the world. I had dozens and dozens of opportunities to kill men who were sworn to kill me and the men and women I served with. I never killed one of them. I was so fucking good at hand-to-hand combat, that I chose time and time again to disarm the enemy fighters, secure their capture and turn them over to my commanding officers. I was never in a confrontation where I had to kill an enemy fighter to save a teammate. That would have been easy. I had to explain many times why I didn't kill the guys. I was embarrassed most of the time, but always made the excuse that I thought the suspect had intel that we could use and killing them would have ended our chances of getting valuable intel." Deklan listened to his own words while keeping one eye on the revolver on the table. Gallardo hadn't moved yet, but he would. Gallardo gained a renewed sense of continued mortality as the combat explanation detailed Deklan's inability or unwillingness to kill people. Good intentions rarely went unrewarded.

"I loved to sit down with my SEAL teammates after a mission or meet at the bars when I was back in country at my San Diego home base. We drank like every night was New Year's Eve and we talked about the missions and the kills. I lied my way through those nights and was certain that my SEAL teammates would have been ashamed of me if they knew that I never killed anyone. I was a professional fighter before I joined the Navy. I beat men half to death for a living. What was my fucking problem? Let me give you an analogy that you may or may not understand." Deklan's blathering had a purpose that would soon play out. Novak continued as if the soliloquy was a personal epiphany.

"In the movie **Braveheart**, a great film, King Edward I of England known as *Longshanks*, invoked the practice of *jus primae noetis (right of the lord)* to breed out the Scottish people. The practice was taking the virginity from peasant women, most often on their wedding night. English noblemen and feudal lords had the right to take women sexually on their wedding night and the women were taken in broad daylight in front of entire villages. The children in the village who witnessed the brutality and barbarism from the English noblemen grew to hate the English more than their parents. The ISIS and Al Qaeda fighters were Islamic fanatics, barbarians to me, like the English noblemen. If I wasted one, then I got one terrorist off the planet, but I created a dozen more. I felt bad because I never killed anyone, until I was asked to intervene for Kaley." Deklan's remarks confused Carmen Gallardo.

There was a short silence. Carmen Gallardo lunged across the table and reached for the revolver that had been sitting in clear view. Gallardo grabbed the gun, but before a shot could be fired, Deklan raised his Glock 17 and fired two shots into Gallardo's chest, (known as a double-tap or Mozambique Drill), and then followed up with a shot to the forehead, that was placed with pinpoint accuracy. The three shots killed Gallardo immediately. The Glock 17 had a suppressor attached. The shots sounded like three short spitting sounds. Gallardo fell backwards and off the chair. The .38 revolver had not been fired. Gallardo's hand was still wrapped around the black revolver. Deklan used semi-jacketed ammunition, and the shot fired into Gallardo's forehead did not exit the skull. Skull shots that had not pierced the skull with an exit wound produced minimal blood loss. The head shot by Deklan Novak entered from the frontal bone and did not penetrate the occipital bone for a clean exit. The floor did not fill with a river of blood.

Deklan calmly stood up and walked to the front door. Deklan palmed the entrance door and pushed it closed slowly but did not lock the door. Deklan put his ear to the door and waited. There were no sounds coming from the hotel hallway. Gallardo was dead on the floor in his hotel room. Deklan Novak walked back to the table and sat down, peering over at Gallardo's dead body.

"I heard this song as a kid maybe a couple thousand times. The song was my mother's favorite song from my mother's favorite band. I am my mother's son. This may not be the Hotel California, but the beast died in the master's chambers." ...and Deklan imagined the lyrics while the song played in his head:

> *"The last thing I remember, you were running for the door. I had to find the passage back to the place I was before. Relax, said the night man. We are programmed to receive. You can check out any time you like, but you can never leave."* *

Novak popped a Skoal Wintergreen Bandit into his mouth. Deklan wondered about bearing the residual psychological responsibility for a combat death versus an execution. He could not imagine that unexpected noises or cars that backfired would eventually trigger any anxiety or irrational zombie-like paranoia inside him. SEALs were taught not to second-guess a combat kill. SEALs had morality and righteousness behind the crosshairs. Wolves with table manners. The first kill took a while, but post deployment therapy was not in Deklan Novak's future.

Deklan stared at the body for another minute or two, then removed his tactical gloves and unscrewed the suppressor on his Glock 17. Novak dialed 911 when the room was sufficiently staged.

*Don Henley, Glenn Frey & Don Felder... music and lyrics by Eagles (not The Eagles). Hotel California: Album release date 12/18/1976. Single release date 2/22/1977.

CHAPTER
TWENTY-SEVEN

LATE FALL 2014...THE CASE FOR SELF-DEFENSE

THE FIRST INTERROGATION took place inside Room 303 at the Hotel Steamboat Springs. Two uniformed officers arrived first, and Detective Ashley Grammer arrived shortly after Deklan made the initial call. A first officer's duty at the scene of a homicide is to execute three primary duties. 1) determine if the victim is alive or dead; 2) apprehend the perpetrator; if possible or give the appropriate notification if the perpetrator is escaping and 3) safeguard the scene and detain witnesses and suspects. Even though detectives and criminal investigators will conduct the formal investigation, the first officer has the responsibility of initiating the investigation and preserving the crime scene as it was upon arrival.

Deklan was the suspect. The victim was dead, and the scene had not been tampered with in any manner since the first officer arrived. Detective Grammer arrived 22 minutes after Deklan called 911 and reported a gunshot victim. The brief emergency call gathered the preliminary information that the shooting was a self-defense incident that resulted from an intruder breaking into the hotel room. The 911 call established that the intruder was dead, and the shooter had called 911. Deklan spoke to Ashley Grammer for approximately 20 minutes at the hotel. The Glock was taken in as evidence, minus the suppressor, which Deklan had disposed of prior to the police arriving. The tactical gloves were placed in Deklan's suitcase at the same time. Deklan called Pat Gabriel and told him he was about to be brought in for questioning. Pat Gabriel had planned for as much. An officer or officers would soon be arriving to bring Pat Gabrial in for questioning.

The Steamboat Springs Police Department shared the same building with the Routt County Sheriff's Department and was located on Shield Drive off Lincoln Avenue, west of the downtown district where Sunpie's and the Eighth Street Steakhouse were located. Pat Gabriel and Deklan Novak were interviewed separately by Steamboat Springs PD and the FBI. The two men met in the lobby after no charges were filed against the former Navy SEALs. There had been many questions that remained, but the case had been boiled down to the following facts:

1. Deklan Novak shot Carmen Gallardo three times after Gallardo broke into Novak's hotel room with a key card, pulled a gun on Deklan Novak and Deklan Novak feared for his life and responded with deadly force. Carmen Gallardo was found inside the hotel room and pronounced dead at the scene by the first officer on the scene.
2. The two men had an altercation at Sunpie's earlier in the evening but otherwise, had not met prior to the Sunpie's altercation.
3. Deklan Novak knew of Carmen Gallardo from his friendship with Behr Thomas, the ex-husband of Gallardo's girlfriend, Raya Nolan Thomas.
4. Behr Thomas and Deklan Novak were in Steamboat Springs last spring to visit Behr's daughter. Deklan Novak had entered the home of Carmen Gallardo six months prior to the shooting and found Kaley Thomas locked in a cage. Child pornography evidence and evidence of factitious disorder imposed on another were found in the mobile

CHAPTER TWENTY-SEVEN

home of Raya Nolan Thomas. Child abuse charges were filed against Gallardo and Kaley's mother only to be dropped months later.

5. Carmen Gallardo never met Novak and did not know of Deklan Novak except from court transcripts. Novak's breach of the Raya Nolan Thomas mobile home was not justified. The claim that Deklan Novak heard cries for help from inside the mobile home could not be verified and any evidence found inside the mobile home was inadmissible evidence and the charges against Raya and Gallardo were dropped.
6. Deklan Novak assumed he dropped his key card at Sunpie's during the physical altercation with Carmen Gallardo.
7. Novak claimed that Gallardo initiated the physical altercation in Sunpie's, a claim disputed by the bartender on duty, but Colton James was reluctant to give any further details other than he thought Novak started an argument.
8. Gallardo's car was in the Hotel Steamboat Springs parking lot at the time Gallardo was shot.
9. A .38 revolver with Gallardo's fingerprints on it was found inside the hotel room next to the body. The gun had been stolen three years before from a vehicle in Edwards, Colorado.
10. Novak claimed Gallardo rushed into his room with the .38 revolver in his hand and aimed the gun at Novak's head. Gallardo was irate about getting his ass kicked earlier that evening. Novak assumed Gallardo had come to kill him. A man broke into Novak's hotel room, pointed a gun in Novak's face and told him he was going to kill Novak. Deklan Novak claimed he had no choice but to defend himself.
11. Novak pulled his own weapon, a Glock 17, and fired three shots at the intruder before the intruder could fire his weapon. Unlikely, but that was the account related to the detective by Deklan Novak.

12. Did Deklan Novak know Carmen Gallardo was at Sunpie's that evening and did Novak go there to initiate a conflict with Gallardo? Why was Deklan Novak in Steamboat Springs? Why did Deklan Novak have a Glock 17 pistol and where did it come from? Deklan Novak claimed that he did not know Gallardo was at Sunpie's. Novak was in Steamboat Springs to meet an old Navy SEAL buddy for two days of skiing. The Glock 17 belonged to Pat Gabriel, Deklan's Navy SEAL buddy from Denver. The Glock 17 was Pat Gabriel's gun, registered and legal. Gabriel and Novak both claimed they were former Navy SEALs, confirmed, and that Navy SEALs traveled with guns. Retired or active Navy SEALs are trained to be experts with small arms and long guns. Gabriel brought an extra pistol to lend to Deklan Novak while in Colorado. Checking a pistol at the airport was not a recommended practice for retired Navy SEALs. Special Operations Command (SOCOM) in San Diego issued retirement recommendations to their retired community. Checking a firearm onto an airplane was not on the Retirement recommended list.

Detective Grammar and the Steamboat Springs Police Chief Reece Morris, a Steamboat Springs native, and a Navy veteran, called the FBI Residence Agency in Glenwood Springs, Colorado. Since the original case against Raya Nolan Thomas and Carmen Gallardo, although dropped, involved internet pornography and possible interstate insurance fraud, Chief Morris did not hesitate to contact the FBI. The FBI determined that a federal nexus was present between alleged darknet pornography, possible interstate insurance fraud, and the yet to be uncovered sexual exploitation that Gallardo may have subjected a minor child to. The Glenwood Springs Residence Agency was somewhat autonomous, within the strict boundaries of the FBI, but the SAIC out of Glenwood Springs wanted to nail Gallardo. The

CHAPTER TWENTY-SEVEN

investigative arms of the Steamboat Springs Police Department and the FBI knew that Gallardo and Thomas had walked based on an exigent home entry issue that fell under a 4th Amendment claim that guaranteed the accused to be free from unreasonable search and seizure. The FBI Colorado Field Office was in Denver, but the FBI had an office in Glenwood Springs because the Interstate 70 passage from Denver to the mountain communities was often cut off by weather. Glenwood Springs offered an alternative to reach certain areas sooner. Special Agent in Charge (SAIC) Laraine Payne and Special Agent Cooper Riley arrived in Steamboat Springs early the next morning. The two FBI agents had called in an Evidence Response Team (ERT) to assist with forensic needs and DNA processing. Surveillance tapes from Sunpie's and the Hotel Steamboat Springs were collected and viewed numerous times. Forensic evidence was collected at the crime scene and from Sunpie's. Carmen Gallardo's prints were the only prints found on the .38 revolver.

The FBI does not send one agent to investigate a federal crime scene. Agents travel in pairs and will call in whatever assistance is necessary. SAIC Laraine Payne was in her early forties, loved working in Colorado and had previously been assigned to the FBI Field Office in Honolulu. SAIC Payne was attractive. At 5'10" tall, with long blonde hair tied back in a ponytail, high cheekbones and a Scandinavian heritage gave her a fashion look that could have come from Vogue. Add a gun to the mix and an FBI shield, Laraine Payne was intimidating and striking at the same time. The tapes were watched and SAIC Payne interviewed Novak and SA Cooper Riley interviewed Pat Gabrial. Payne and Riley reviewed their notes from the morning sessions. The tapes showed Gallardo at the hotel, in the parking lot moving to the main entrance at the time Novak claimed. The hotel hallway tape showed Gallardo leaving the elevator and moving towards Novak's room. Gallardo was shot at close range from the front as the two men faced each other. These facts supported a self-defense claim. Three shots to the victim's back would have made self-defense a much taller tale to spin. The tapes from Sunpie's did not show much. The two men were visible, and their altercation was grainy on tape. The audio from the encounter did not exist. There was no way to determine who initiated the fight from the tape. Payne called the shooting self-defense but suspected the Charles Bronson (1974) or the Bruce Willis (2018) character from Death Wish, Paul Kersey might have had some influence on the events that occurred the night before and led to the demise of Carmen Gallardo. The state of Colorado had become an infinitely better place to live following Gallardo's expiry.

Detective Askley Grammer had her own theories as to what happened at Sunpie's Bar and what happened at the Hotel Steamboat Springs on the night of the shooting. Detective Grammer released Novak and Gabriel after Grammar had a meeting with the FBI agents in town. Grammar asked Deklan and Pat Gabrial to stick around town for a few days in case any further questions arose. Both men informed Detective Grammer that they had to be back in Denver and Chicago within the next two days and would keep their plans to leave the area. They would remain available for any further questioning or interviews, but the contacts would come from Chicago and Denver. Neither man was charged, and neither man was running away from the case. Pat Gabriel and Deklan Novak met in the lobby of the police department building and walked outside. The morning sun was high in the sky.

"I'm sorry I got you into this, Pat." Deklan began.

"No, you're not." Gabriel countered.

"No, I'm not. You're right." Deklan laughed as the two men shared a locked-thumb handshake and a man hug.

"I'll be back in Denver tonight. Keep me posted." Pat Gabriel said good-bye and added. "There aren't many of us around, so who loves you, bro?"

"Safe travels. I'll make sure they get your gun back to you." Deklan promised.

"Cool. No worries. I've got more. Later, Deklan."

Navy SEALs lived by habits learned as a SEAL. Loyalty, putting others before oneself, being organized, reflective, detail-orientated, and never getting comfortable...were the credos that bonded Pat and

CHAPTER TWENTY-SEVEN

Deklan as they parted ways in Steamboat Springs. Deklan had school and football practice to get back to, not to mention four boys and a wife that had become strangers in his world. The first kill of Deklan's unorthodox journey occurred inside a relic hotel room from the 70's. Deklan was not a candidate for a combat medal. Regret, remorse, and any sliver of guilt had not invaded Deklan's space since the incident unfolded as planned.

Deklan returned to practice on Monday after classes. Behr Thomas called Deklan into his office after practice. Deklan arrived sweaty and partially still in uniform. The shoulder pads and practice jersey had been shed. Deklan's spectacular full-sleeve tattoos painted a muscular landscape from a sleeveless Navy t-shirt. The football pants were stained with frozen mud, winter grass and a nasty combination of oral excretions that defied a description. Deklan stood at the desk of his defensive coordinator.

"Tell me about Steamboat Springs." Behr instructed his former traveling partner.

CHAPTER TWENTY-EIGHT

LATE SPRING 2015...OFFICE OF THE PRESIDENT, NORTHWESTERN UNIVERSITY: TAYLOR CROWNE

THE MEETING WAS WELL-ATTENDED in May 2015. The summit was held inside the office conference room adjacent to the Office of the President at Northwestern University. President Taylor Crowne called in all the people involved. In addition to Crowne, Northwestern University Board Secretary and Advisor to the President Anna Strickland was in attendance along with Vice-President for Student Affairs Kaitlin Wilson, Athletic Director Jim Tunney, Assistant AD Manny Griffith, NU Head Football Coach Lance Adams, Defensive Coordinator Behr Thomas, Offensive Coordinator Chase Parker, Special Agent in Charge (SAIC) Donald W. Meeker Jr. from the Chicago FBI Field Office at 2111 W. Roosevelt Road in Chicago and Chicago FBI Special Agent Julian Bavetta. The meeting, arranged by Northwestern University President Crowne, was the result of a call from the J. Edgar Hoover Building in Washington D.C. The call came from the Director of the Federal Bureau of Investigation, Mark Beatty. The last person to attend the meeting was Deklan Novak.

Deklan Novak was worried. The 2015 spring football game and spring practices were behind him. The upcoming year was going to be the best chance for Deklan to begin the fall campaign as the starting Mike linebacker for the Wildcats. Deklan envisioned the 2015 season as a legitimate chance for the Wildcats to compete for a national championship with an aging, yet effective defensive Co-Captain. The meeting that had been hastily arranged in Taylor Crowne's office was beyond troublesome. Deklan walked into the conference room and the confusion grew after the conference room grew oddly quiet following Novak's appearance. What had the FBI found regarding the Gallardo case? Was this the end of an abbreviated college football career? Was a murder charge pending? Were there uniformed law enforcement officers waiting somewhere in the building to take Deklan into custody. There would be no other reason to gather such an illustrious panel of spectators. The meeting had to be the first step to exorcise from the university, the ex-Navy SEAL, officially and unceremoniously. The panel had to have been assembled to bear witness to the fall from grace and to be present when a federal arrest warrant from Colorado was read aloud.

The table was huge. Deklan thought it had to have been 18-20 feet long. The table was made of polished cherry wood with a Sapele band highlighted by black and turquoise inlay lines. The table was able to seat 16 people. The attendance in the room was 11 people. NU President Taylor Crowne sat at the head of the table. He was reviewing some notes and spoke to FBI Chicago SAIC Donald Wheeler to his immediate right on more than one occasion before he began the meeting. Taylor Crowne carried with him a remarkable resume. Crowne took over the job at Northwestern as President roughly fifteen months ago in the winter of 2014. Before NU, Crowne was the President of the University of Montana, dean of the University of Arizona College of Law, and the dean of Washington University School of Law. Deklan was a bit intimidated by the panel. A review of the exit route possibilities was under way. Deklan tried to calculate how to remove himself from the premises quickly and unexpectedly. President Crowne stood up to address the room.

"Greetings," Crowne began with starched dignity. "When the school was approached by the FBI, I must admit, I was more than intrigued. Deklan Novak has brought an unusual level of interest and integrity to Northwestern University. I was contacted by Special Agent in Charge Donald Meeker from the FBI Chicago field office two days ago. SAIC Meeker brought me up to speed on the Bureau's plans after I explained what FBI Director Mark Beatty and I had discussed on a phone call earlier in the week. I had been unaware of the circumstances that unfolded in Steamboat Springs last year. I was heartbroken to hear what Coach Thomas' daughter had been subjected to and I was very happy to find out that ultimately, Kaley and Coach Thomas have been reunited and are living together here." Taylor Crowne scanned the room and stopped when he was looking directly at Deklan Novak. Deklan's heart raced and his adrenaline began to rise dramatically. The meeting had to have been called to address the shooting at the Hotel Steamboat Springs. Crowne was already discussing the circumstances in Steamboat. The brief silence was designed to be uncomfortable, Deklan imagined, as it was choreographed by a seasoned attorney.

"I wish I had been advised of the travel to Steamboat Springs by Behr Thomas and Deklan Novak, but I understand why I was not advised. I read with great consternation about the charges leveled and eventually dropped against Ms. Nolan Thomas and Mr. Gallardo. I was relieved to read that Mr. Novak was unharmed by the intruder into his hotel room. Coach Adams informed the President's office after he became aware of the police report that outlined a self-defense case based on the state laws regarding deadly force and the use of a firearm. The details of the evening were complicated to say the least, but the FBI came to me almost one year after Mr. Gallardo met his untimely demise. They presented new evidence, briefed me on a new course of action, and put the gears in motion to end up where we are today. Special Agent Meeker, I give you the floor." Taylor Crowne sat down.

SAIC Donald W. Meeker Jr. was 45 years old. Donald Meeker was from Des Moines, Iowa and stood 6'2" tall and weighed a tight 202 pounds. As with most male FBI agents, Donald Meeker sported a close-cropped Marine-styled haircut and wore a conservatively cut Navy blue suit, white shirt, and a solid-colored red tie.

Donald Meeker Jr. had been assigned to the Chicago FBI field office in 2005 and was appointed SAIC in 2011. In major city field offices, there was an Assistant Director in Charge (ADIC) that was above the SAIC. In Chicago, the ADIC was Monica Corcoran, the second highest ranking female in the FBI at the time she was appointed to the Chicago field office. SAIC Donald Meeker led task force local investigations over the years into dozens of Chicago Outfit mobsters including, Outfit boss Albert Tocco, mobster Nickolas Calabrese, Outfit enforcer and loan shark Frank Calabrese Sr. The Outfit arrests were among the many cases spearheaded by the Chicago field office of the FBI. The Chicago FBI office was on Roosevelt Road, having moved from the Dirksen Federal Building where the field office was located from 1964-2006

"Greetings to all here today." SAIC Meeker began the meeting after he was introduced by NU President Taylor Crowne. "I am not a public speaker by any stretch of the imagination and the meeting here to address an unprecedented action taken by the FBI was arranged by our Director in Washington, D.C. I humbly accepted the assignment, but that did not improve my oration skills in front of a distinguished panel such as yourselves." Meeker took a quick view of the panel at the large table and went on.

"Recruitment at the FBI is not a proactive enterprise that has designated staff members assigned to target individuals that may display the character and the qualities that the FBI has sought from its inception. In fact, I cannot remember one single individual that the FBI Director, the Deputy Director, the Director of Intelligence, the Associate Directors, or any of the Special Agents in Charge from the 56 FBI field offices across the country has singled out specifically to be recruited by the FBI. I do not recall anyone in the Bureau assigned to interview an individual target for the purpose of adding the individual to the organization as a Special Agent. That is, until FBI Director Mark Beatty called the Chicago field office and asked ADIC Corcoran to speak to Northwestern University about Deklan Novak.

CHAPTER TWENTY-EIGHT

President Crowne and Director Beatty had already spoken by telephone about Deklan Novak." SAIC Donald Meeker paused his remarks and turned to face Deklan Novak. The panel at the table did the same. Deklan Novak sat stunned and thought about asking the speaker to repeat himself.

"You see." Meeker continued. "We, at the FBI, have guidelines that prospective agents must meet before they are allowed to apply to the Bureau. The applicant must be at least 23 years-old and no applicant can apply to the FBI after they turn 36 years old, unless they have a veteran's status or federal law enforcement experience. Deklan is a decorated veteran, so his age is not a factor in the Bureau's efforts here today. The Bureau's mandatory retirement age of 57 can also be extended by prior military and law enforcement experience. Deklan, we have vetted you prior to the meeting here. Citizenship in good standing and our assumption that your ability to pass the physical requirements will be a forgone conclusion, the FBI will extend a rare, if not unprecedented invitation, to you to become an FBI Special Agent. The invitation will begin after you graduate from Northwestern University in 2016. The exemplary way you have lived your life and the service to your country has brought your story to Washington D.C. and the attention of FBI Director Mark Beatty. The University was brought into the invitation process because obtaining a bachelor's degree is a requirement to join the FBI. The publicity surrounding your participation in the football program, as an NU student and your former status as a United States Navy SEAL has not gone under the radar. The story has beneficial repercussions for the university and the FBI. The bottom line, however, is that the Bureau has never extended a targeted invitation to join the ranks. As the spokesman for the FBI, we extend an invitation to apply to the Federal Bureau of Investigation to Deklan Novak upon his graduation from Northwestern University in 2016 or whenever that graduation takes place. Deklan, you will then go through the standard application process, interviews, and physical fitness requirements. Upon a successful journey through the application process, you will enter the FBI Academy in Quantico, Virginia to complete the 18-week training regimen that is required to become an FBI Special Agent. We will discuss the process further following the announcement here today and you can take as much time as you need to decide if law enforcement and the FBI is the right path for you. Thank you, Deklan, for the example that you have unintentionally set and the high bar you have established to achieve the unachievable." SAIC Meeker walked over to Deklan Novak and extended his hand. The two men shook firmly and Meeker's outward demeanor spoke of anything but respect and admiration.

The unusual announcement at the University President's office took the participants by surprise. The publicity surrounding the invitation had potentially far-reaching tentacles that boosted the football program recruiting efforts, the university exposure in general, the Federal Bureau of Investigation recruitment numbers, the Navy recruiting offices, the SEAL community, and the military community in general. The Department of Defense recently published the retention and recruiting numbers for fiscal year 2014. Over the past ten-plus years, recruitment and retention numbers have generally met the goal numbers set by the various branches of the military, but the government wanted to exceed goal numbers. The Department of Defense was expected to open all military occupational specialties to women in the next year. The first women to graduate from Army Ranger School took place last year. Green Berets and Navy SEAL female candidates being accepted into SEAL BUD/S training and into the Green Beret Special Forces Qualification Course were under command review and apparently not far behind the precedent set by the Army Ranger School's female graduates. SAIC Meeker and Deklan Novak met in a private office after the conference room meeting ended. Deklan's coaches gave him his due in typical jock format before the private Meeker meeting.

"Special Agent Novak, sir." Head Coach Lance Adams began as they shook hands in the hallway outside the conference room. "Congrats, Deklan. You know that the FBI kept files on John Lennon, Bob Dylan, Michael Jackson, and Mickey Mantle. Can you make sure they don't have one on me? I've been thinking that they have me under

CHAPTER TWENTY-EIGHT

surveillance since I always see a helicopter flying overhead when I get in my car at home." Lance Adams winked as he headed back to his office.

"What do you call an alligator who joins the FBI?" Offensive coordinator Chase Parker asked Deklan. Deklan shrugged his shoulders.

"I don't know." Deklan replied.

"An investi-gator." Parker shot back and patted Novak on the shoulder as they both feigned a hysterical sarcastic response. Deklan bowed and shook his head slowly from side to side, expressing the collective embarrassment from the football community, law enforcement, students, and comedians everywhere for the lame joke.

Defensive coordinator Behr Thomas walked up to Novak and gave him a bearhug. Behr whispered a response to his Co-Captain and starting Mike linebacker.

"Holy shit, dude." Behr smiled and he joined the others as they headed back to the athletic offices they occupied. Deklan and SAIC Donald Meeker headed for a private office on the floor.

SAIC Meeker opened the office door abruptly and walked to the windows overlooking the lakefront campus. Deklan followed Meeker into the office.

"Can you close the door, please." Meeker asked curtly. Deklan closed the door.

"This charade is not from our office." SAIC Meeker began. Meeker wanted a private meeting with Deklan Novak ever since he heard about the unusual assignment. "I serve the Bureau with blind devotion, but I do not always agree with Washington. When the FBI Director calls your field office personally and assigns you a task, you thank him and question nothing. I'm sure you know what I am talking about since you were a Navy SEAL."

"Yes, sir. I do know what you are talking about." Deklan replied respectively.

"While I would never outwardly question my superiors, especially the FBI Director, I think this is horseshit. I was told that you will be assigned to the Chicago FBI field office after graduation from the FBI Academy. I was told further that you will be assigned to our VCAC

(Violent Crimes Against Children) squad immediately. The Chicago FBI office, my office, oversees the highest percentage of violent crimes against children per capita in the nation. Why is Chicago such a hot bed for child abuse, child murders and child kidnapping when the city is ranked 42 out of 100 in overall crime statistics measuring rates per 100,000 people.? We do not have an answer for that question. What I do know is that an assignment to the VCAC squad must be earned after a two-year probation and a few years on the street as an agent. The Director is bypassing those details in your case. I have a problem with that, but my issues will not stop me from following my orders from the FBI Director and my ADIC." Meeker stopped.

"Yes, sir." Deklan responded and was searching for a clue to what the acronym A-dick meant.

"No one is given their FBI assignment location until week six into the Academy. New agents are sent where we need people, not to their hometowns. I do not know what the basis was for the actions now put into place, but all I see from you guys is more books. Navy SEALs have published more books than Danielle Steel." Meeker remarked.

"Yes, sir." Deklan answered again. "I have no clue who Danielle Steel is, sir." Novak added sarcastically. He did know who Steel was but had never read one of her books.

"I will not be a warm and fuzzy supervisor when you get to my field office." SAIC Donald Meeker moved towards the door and added one more thought. "I read the Glenwood Springs FBI office report on your incidents in Steamboat Springs. If you think you are going to operate like that on my VCAC squad, you are mistaken. That staged break-in at the Hotel Steamboat Springs allowed you to play vigilante hero and walk scot-free. That charade may have even been the reason Director Beatty wants you in the Bureau. I can only guess, but we operate by the book in the FBI. You won't last in our organization, and I am guessing, you will not make it through the academy." SAIC Meeker was not referring to the physical challenges in the academy. SAIC Meeker was leaving but waited for Novak's response.

"You came to me, remember." Deklan replied. "With all due respect, sir, I am certain that I will survive the FBI Academy. I'll get

CHAPTER TWENTY-EIGHT

back to the FBI as soon as I decide if my career path brings me your way." Deklan nodded and did not smile or attempt any further engagement.

Although SAIC Donald Meeker and Deklan Novak were not about to engage in a physical conflict, the body language from Donald Meeker suggested otherwise. Meeker invaded Deklan's space by stepping forward as he talked. Meeker's posture was an attempt to overcome the obvious difference in stature between the two men, the physical and subliminal variance was palpable. Deklan was twice as wide as Meeker, a much stronger physical presence. Novak was a former Navy SEAL, a symbol of instant respect, an immediate erosion of self-confidence in Meeker. Novak saw it often in other men. The observations were not born from an inflated ego, the reverence was real. In Meeker's case, reverence was resentment.

Soft, smooth, and small hands meant that Meeker avoided confrontational physicality. The verbal boundaries surrounding the exchange with Novak gave Meeker a false sense of bravado, while the metaphorical expression based on dogs, the common idiom of "all bark and no bite" rang loud in the room. Deklan Novak decided in that private office and within that moment, to become an FBI agent. Deklan remained silent as Meeker pondered a good exit line. None came to mind and SAIC Donald Meeker walked away.

CHAPTER TWENTY-NINE

1991 & 2017...POINT BREAK: "UTAH, GET ME TWO!" AND THE FBI INTERVIEWS

DURING THE SUMMER before fifth grade, Deklan went to see a new movie, released on July 12, 1991. Deklan went to the movies with a couple of buddies. The movie was Point Break, starring Patrick Swayze as Bodhi, Keanu Reeves as undercover FBI agent Johnny Utah, Lori Petty as Tyler, and Gary Busse as FBI agent Angelo Pappas. The film was directed by Kathryn Bigelow who would later direct iconic films like **The Hurt Locker** and **Zero Dark Thirty**. Deklan was absolutely mesmerized by the film and many of the iconic quotes from the movie found an indelible place in Deklan's nine-year-old mind forever:

Bodhi: "If you want the ultimate, you've got to be willing to pay the ultimate price. It's not tragic to die doing what you love."

Bodhi: "Fear causes hesitation, and hesitation will cause your worst fears to come true."

FBI Agent Angelo Pappas: "Utah, get me two!"

FBI Agent Johnny Utah: "Wars of religion always make me laugh because basically, you're fighting over who has the best imaginary friend."

Bodhi: "It's basic dog psychology. If you scare them and get them peeing down their legs, they submit. But if you project weakness, that promotes violence, and that's how people get hurt."

FBI Agent Johnny Utah: "You're sayin' the FBI's gonna pay me to surf?"

FBI Agent Johnny Utah: "Vaya con Dios."

When Deklan Novak watched Point Break, his career plan was set in stone. Deklan Novak was going to be an FBI agent, period, end of story.

Deklan got distracted as he went through high school and eventually found a Navy SEAL poster inside a Navy recruiting office that officially changed his career path. There was the trip to Annapolis in 1989 that ignited a Navy spark inside Deklan. Then came Point Break. For many years after the fourth grade, after the 9/11 attack on New York City, and after a decorated tenure as a United States Navy SEAL, Deklan hadn't thought much about the FBI until he had a chance meeting with the Chicago Field Office Special Agent in Charge, Donald Meeker.

The FBI does not recruit individuals. The FBI Special Agent is a federal position. There are no wine and dine meetings to catch a rising star. Occasionally, an active FBI agent will assist a friend or a former colleague to apply to the FBI. In Deklan Novak's case, the FBI Director intervened, and all bets were off. Deklan had a clear path to the Chicago field office, but that path did not eliminate the normal application process that had to be cleared and passed by Deklan Novak. The First Phase written tests and the initial interviews were held in a hotel conference room across the street from the Chicago Field Office. Deklan passed the math and essay portions without a problem. Since the FBI required a college bachelor's degree as a minimum requirement to apply as an FBI Special Agent, the onus of the interview process had little to do with each applicant's college degree. They all had a college degree, so the hiring personnel focused on what the applicants had done in other phases of their lives.

The Second Phase interviews were background intensive and scenario-based sessions held before a three-person panel. There were

CHAPTER TWENTY-NINE

seven applicants on the day Deklan Novak arrived for Phase Two interviews. Deklan was able to detail his background during the Second Phase interview. When Deklan recounted a brief career as a professional fighter, his decade of service to the nation as a Navy SEAL, multiple combat deployments and a walk-on starting linebacker position at a D-1 school, the ongoing Second Phase process was pretty much a done deal. Deklan received a formal offer for employment in the FBI by mail. The next phase was the physical fitness test that every incoming applicant had to pass to attend the FBI Academy in Quantico, Va.

The Physical Fitness Test (PFT) for Deklan Novak was held at the Theodore Roosevelt High School track in Chicago, Illinois. The physical fitness test for the FBI included a 300-yard sprint, a 1.5-mile run, pushups, sit-ups and pull-ups. Deklan assumed he would breeze through the PFT. The sprint was easy. The 1.5-mile run was easy.

Deklan was often asked to compare BUD/S with the FBI Academy. Without trying to denigrate the FBI Academy in any manner, Deklan compared Quantico to a Catholic summer camp when asked to define the difference between Navy SEAL training and training for the FBI. Deklan always emphasized that the physical comparisons were not close, but the FBI did a fantastic job in preparing new agents for the life and death decisions that they would be asked to make on the street every day of their new careers.

During BUD/S training, Deklan was able to knock out hundreds of pushups daily and on most days in Coronado, the candidates were asked to do just that. BUD/S candidates were ordered to do thousands of pushups on the grinder in Coronado during the First Phase of SEAL training. At Theodore Roosevelt High School in 2017, Deklan began his pushups to pass the FBI required calisthenics necessary to become a Special Agent. Applicants were required to knock out a minimum of 70 pushups in the allotted time, a slam-dunk for a SEAL. When Deklan started his pushup test, the instructor began counting quickly. The elbows had to reach 90 degrees, and the applicant could not stop his or her motion during the test. Deklan's instructor began 1, 2, 3, 4, 5, 6…. etc. Suddenly, the instructor started repeating numbers. 28, 28, 28, 29, 30. Deklan didn't understand, but kept up his remarkable endurance and show of strength. When the time had ended, Deklan had completed 77 pushups. Deklan was shocked. He had counted more than 110. The instructor informed Deklan that the FBI pushup is different from the Navy SEAL pushup. Applicants were not allowed to pause for a split second at the top or bottom of each pushup, a distinct difference from the pushups required as a Navy SEAL or anywhere else on the planet for that matter. Deklan asked to retake the pushup portion, which he had passed regardless of the FBI pushup violations. The request was granted, and Deklan knocked out 122 pushups in the allotted time, then glared at the pushup Nazi who had apparently enjoyed the encounter.

CHAPTER THIRTY

2017...FBI ACADEMY QUANTICO, VA. (18 WEEKS)

MOST NEW AGENTS arrived by bus on the Sunday afternoon before the official start on Monday morning. Deklan Novak drove to Quantico from Chicago. Arriving on I-95 at Exit 148, Russell Road, Deklan drove the 2015 Ford F-150 Raptor that he purchased after his divorce was final. Deklan had been frugal during his SEAL years and had received a healthy inheritance when his grandfather, Goran Novak, passed. The inheritance was not subject to the divorce settlement as the money was considered separate property belonging exclusively to the beneficiary and not subject to be divided in a divorce settlement. The 2015 Ford F-150 Raptor had a retail price of $55,000 plus. The Raptor rolled on 34" 295/70/17 jet-black tires, offroad GF2 wheels and was a high-performance off-road pickup truck equipped with a 6.2-liter V-8 engine that housed a 411-horsepower rocket ship under the hood. The royal blue rolling tank had black leather seats, a black front cowl hood, a rear black roll bar with five LED lights resting across the top of the cab. Russell Road curled around a wooded area off I-95 for approximately three miles and eventually led to the main gate at the Quantico Marine Base.

The massive base was home to the Marine Corps Combat Development Group, Marine Corps Recruiting Command, Marine Corps Systems Command, Marine Corps Officer Candidate School, and as many as thirty Marine Corps Commands, Schools, and Squadrons. The base was also home to The FBI Training Academy, The FBI National Crime Lab, The DEA Training Academy, Naval Criminal Investigative Service, and The National Academy (a ten-week law enforcement training course for established law enforcement officers and agents held in Quantico four times each year). Quantico is not a location for the Marine Corps recruit boot camps. The United States Marine Corps Recruit Training (boot camp) was held at one of two locations, Paris Island, South Carolina and San Diego, California.

The FBI National Academy shared the main training facility with the FBI New Agent Training Course or the FBI Basic Field Training Course (BFTC). The dorms for New Agent Trainees (NATs) were split between the divisions as the agents entered the main FBI Academy building. Madison Dormitory was where National Academy students stayed during their ten-week course. National Academy students were assigned single person rooms and private bathrooms. Most National Academy students were seasoned law enforcement officers who waited for an opening to attend the National Academy, and they had been sent by their own local police departments. The National Academy hosted delegations of law enforcement officers from around the world as part of each session. Washington Dormitory was the New Agent Trainee dormitory. New agents were housed in college-like dormitory settings, two to a room and a shared bathroom for every two rooms.

The entrance to the base at Quantico had multiple security layers and Deklan eventually parked the Raptor in the parking lot next to the main dormitory buildings. Deklan entered through the main doorway under the FBI Academy sign that read Fidelity, Bravery, Integrity…the hallmark credo of the FBI and the origin of the acronym namesake. Deklan walked past the 9/11 Memorial, a MLK Memorial wall, and a wall honoring the National Academy and the origin of the FBI in 1908. The dorms were seven stories tall. Deklan took the stairs up to his assigned room. New agent trainees (NATs) were prohibited from using the elevators until they passed the Physical Fitness Test (PFT) after the first week of the BFTC. New agents were required to pass the PFT before their applications were approved and each new agent trainee had to pass the same PFT in Quantico during the first days of the Academy. Deklan's roommate was already unpacked and settled. There was no one in the room when Deklan arrived. Deklan recalled his first day in Kenosha at Northwestern football camp and the extensive electronic gear each player brought with them to the University of Wisconsin – Parkside Campus. The NATs in the FBI dorm rooms were not allowed to bring gaming devices, audio music systems, Bluetooth speakers, coffee makers or microwaves. There was one television per room, in between the two desks, provided by the FBI. NATs were allowed to bring books, cell phones, laptops, and pictures. It was late Sunday afternoon. The BFTC started at 6 a.m. the next morning.

Deklan walked down to the cafeteria. New Agent Trainees were milling around over meals. There was a nervous energy in the room, one of anxiety and excitement. Deklan was neither excited nor nervous. The BFTC was going to be Deklan's fifth initiation dance, and the excitement level was a bit reserved. There had first been the informal, yet brutal MMA rookie training before the UFC would allow a fighter to book a sanctioned fight. There was Navy boot camp at Great Lakes followed by BUD/S training in Coronado. Once a Navy SEAL was assigned to a SEAL Team, there was an extensive period of training with the new team before a SEAL was operational. The last training facility Deklan went through was the Northwestern football camp at UWP. The Northwestern camp included the rookie hazing rituals and a difficult summer camp training regimen but nothing remotely close to BUD/S training. The FBI curriculum did not intimidate the former Navy SEAL.

Deklan was not hungry as he walked through the cafeteria dining hall. The dining hall was busy and loud. The incoming class of NATs had heard about the former Navy SEAL and the business in Colorado. The case had attracted some national attention publicly but was common knowledge within the Academy. Heads turned almost on cue as Deklan made his way through the room. Deklan wore a UDT/Navy SEAL Instructor tee-shirt, creased Wrangler jeans, and cowboy boots. Deklan's age stood out. Deklan's three-day salt and pepper stubble would be gone by morning. The tattoos stood out and the physique was an eye-opening reality-check as the new agents measured up the new competition. The faces inside the FBI Training Academy cafeteria did not look confident as they eyeballed the modern version of the original UDT Frogman.

Deklan walked to the Boardroom, an adjacent bar with beer, wine, and liquor. The Boardroom was usually used by National Academy law enforcement session students. The FBI preferred the older, NA veteran police officer students to stay on campus, rather than go driving into town looking for a bar. The Boardroom was not off-limits to NATs, but few ever rambled in during the BFTC 18-week timeline, and no one from a new agent class went to the Boardroom on the first night. Deklan sat at the bar and ordered a club soda with a lime.

"Are you a new agent?" A younger man asked Deklan. Deklan turned to see a young man with his Polo shirt buttoned all the way up, often a sign of uptight formality or unnecessary stiffness. The new agent addressing Novak wore the uniform issued after the four-day One Program for all NATs and intelligence officer trainees. The One Program began on Monday morning and taught the incoming employees about the FBI's structure, mission, and culture. The uniforms required and issued to NATs after the on boarding One Program had concluded consisted of a navy-blue Polo shirt, color-coded lanyard, a holstered red Glock 19 non-actionable pistol, and tan khaki pants. Obviously, the younger new agent had studied the NAT photos prior to his arrival in Quantico.

"Yes." Deklan answered and sipped his club soda.

"It is generally frowned upon for a new agent to be in the Boardroom, especially on day one. We are judged as one unit, and your actions reflect on us all." The self-appointed training academy hall monitor pointed out.

CHAPTER THIRTY

"It is not the first day. Tomorrow is the first day." Deklan mentioned. "What is your name?" Deklan asked.

"My name is Jordan Dulles, like the airport." Jordan Dulles answered nervously.

"If new agents are generally frowned upon by being in the Boardroom," Deklan began his inquiry. "Then why are you here?"

"Excuse me?" Dulles stumbled.

"OK, Jordan Dulles, like the airport." Deklan replied. "Just fucking with you, my man. However, you are officially not allowed to speak to me again. Ever."

Deklan stared at Jordan Dulles briefly while Dulles stumbled for a response that did not come. Jordan was a Caucasian man, tall and thin, with a buzz cut, a cleft chin and sunken cheeks. Jordan Dulles might have been mistaken for a young, thinner Rowan Atkinson starring in the 1997 film, Bean. Jordan did not reply and gingerly walked away, looking back once, but Deklan had returned to his club soda. Jordan reviewed his purpose and the challenging task to become an FBI Agent that would begin on Monday morning. The hall monitor responsibilities were abandoned as Jordan Dulles returned to his room, hoping no one had witnessed the encounter with the former Navy SEAL. Jordan Dulles stayed awake that night for a time and prayed that Deklan Novak was not assigned to his New Agent Trainee section. The incoming class of 200 NATs were divided alphabetically into four sections of 50 students each…Alpha, Bravo, Charlie, and Delta. Jordan finally fell asleep after he had assumed Novak was sufficiently down from Dulles and therefore, the former Navy SEAL had to be in a different trainee section.

After the One Program was completed and each new agent trainee was sworn in, the NATs began their new agent boot camp. The normal days during the first six weeks of the FBI Academy were structured as follows:

6:00 a.m. – 7:15 a.m. was dedicated to power physical training (push-ups, sit-ups, pull-ups, sprints, 1.5-mile runs)

7:30 a.m. - Breakfast

8:00 a.m. – 11:45 a.m. legal classes

Noon – Lunch

1:00 p.m. – 4:30 p.m. Gun Range Training with the standard issued FBI sidearm Glock 19M, 9mm pistols with a 17-round magazine capacity, a Remington 870P 12-gauge shotgun, and a Colt Pattern Carbine Model LE6920 Series assault rifle, with a wide variety of 5.56 mm ammunition available, featuring a 30-round magazine capacity, Troy Flip-Up Front and Rear Sights, (*an AR-15 variant... a similar weapon with an M203 grenade launcher attached was fired by Al Pacino's character Tony Montana in the 1983 film, Scarface during the final scene: a.k.a. say hello to my little friend*)

The first six-weeks of the FBI Academy had NATs firing 4,000 Glock rounds, 120-150 Remington rounds, and 600-700 Colt pattern Carbine rounds during the training sessions.

The first day on the range at the FBI Academy, Deklan had a wake-up call. Many NATs had military experience. Many NATs did not have any military training, and some NATs had never touched a firearm in their lives. The range instructors at the academy took the same approach regarding gun operations and safety with every NAT on the first day at the gun range. The instructors demonstrated the proper and safe handling of a firearm by holding the weapons with the new agents and assisting on the trigger-pull as the new agents fired their weapons for the first time on the range. This meant that an FBI Academy instructor held Deklan's Glock 19M with him and helped him pull the trigger as the NATs aimed at the targets for the first time.

"You are aware that I served as a Navy SEAL for almost eleven years, that I had multiple combat deployments to Afghanistan and Iraq, that during BUD/S Phase Two and Three training in Coronado, we fired at least 100,000 rounds during training on 9 mm pistols, and that as a Breacher for SEAL Team One, I was responsible for securing a tactical advantage for our sixteen man teams as we gained entry to enemy positions and secure locations? Can we skip the handholding on the weapon now that you do know that I have an extensive background on all the weapons we will be shooting here at the academy? I appreciate that you want to help me hold onto the trigger and show me how to fire a gun, but there is nothing personal here, I'm good on

CHAPTER THIRTY

my own." Deklan asked with the utmost respect and reverence. Deklan also smiled slightly as he waited for his instructor's response.

"I am very aware of your background Agent Novak." The veteran FBI Academy instructor commented. "The entire NAT class is well-aware of the Navy SEAL joining our ranks. However, we are required to follow the academy guidelines for new agent training on the range, especially on the first day. Now, go along with the nonsense for today and hope that they do not require the instructors to wrap both arms around the new agents during assault rifle training. We could potentially be spooning for our carbine training, so let's get through today. Any further touching will require you to buy me dinner first."

"Let's get through today." Deklan agreed as he smiled and eventually burst out laughing. Deklan and his instructor got along great after the former Navy SEAL learned on Day One that in the FBI Academy, he was there on their terms only. The first six-week range shooting was flat range shooting with alternating single hands from distances of 3 and 5 yards. For distances beyond 5 yards up to 25 yards, shooting exercises were conducted with two hands on the pistol. Deklan put on a show at the firing range on day one. NATs were required to always wear a firearm during the day. At the range, the NATs used real weapons. On campus, the NATs carried red Glock 19M non-actionable pistols. During the Hogan's Alley training, the NATs carried blue sidearms with non-lethal, 10-round magazines.

The major difference between SEAL training and the FBI Academy was paramount to Deklan understanding his new role. At BUD/S, the SEAL instructors sought to eliminate the weaker candidates, and they accomplished their task successfully as evidenced by the 80% failure rate at BUD/S. The mantra for BUD/S training was **"The Only Easy Day Was Yesterday**." BUD/S instructors were the gatekeepers to the SEAL community. The training at BUD/S was designed to test every aspect of a candidate's mental and physical endurance to such an extreme that only a select few candidates were able to qualify to become Navy SEALs. BUD/S training was designed to select only the ones who wanted it more than anything in life.

At the FBI Academy, the candidates had already been selected. The potential new agents had survived a rigorous selection and hiring process that resulted in most candidates being turned away. Once the candidate had been selected to attend the FBI Academy, the FBI did not want to drive the candidates to failure. The FBI Academy was about training new agents to be efficient, safe, effective, tough, smart, adaptive, resilient, and the FBI Academy first and foremost, wanted their candidates to graduate.

CHAPTER
THIRTY-ONE

2017... FBI BASIC FIELD TRAINING COURSE (BFTC), FIREARMS TRAINING, LEGAL, DEFENSIVE TACTICS, HOGAN'S ALLEY AND A QUANTICO GRADUATION CEREMONY

DEKLAN PASSED the PFT again on day two in Quantico without the pushup Nazi present. One of the surprising aspects of the FBI Academy to Deklan was that only the new agents who failed the PFT were required to train daily. These new agents were given ample opportunity to retake the PFT until they passed. Since the FBI had already selected these new agents, they did not want to see anyone fail the physical requirements or DOR (Drop on Request). Deklan Novak carved out a couple hours each evening to train on his own...whether it was running outside or on the indoor track. Deklan spent many hours in the weight room assigned for the BFTC new agents while completing copious numbers of calisthenics (pushups, sit-ups and pull-ups). Deklan's training sessions were not executed to impress anyone. Over the years, Deklan had maintained a level of fitness that he wore well and that took many hours each week to maintain. The sessions became gossip fodder for the academy. Some resented the efforts, assuming the sessions were designed to show up the class, but others asked to join in. Deklan gladly brought anyone along for the ride.

The FBI Academy weekly training schedule included the following sequence:

Week 1...One Program and Day One Oath

Week 2-5...Breakfast, morning legal classes 8:00 a.m. – 11:45 a.m., lunch, gun range 1:00 p.m.-4:30 p.m., dinner

Week 6...Orders night (where NATs found out where they were being assigned)

Week 6-16...Hogan's Alley: Tactical shooting, gun range, fight club, deadly force training, real-life action scenarios, firearms, judgement shooting, defensive tactics, Capstone (a 6.1 mile run through a hilly, wooded terrain built by Marines,), Tactical Emergency Vehicle Operations Center (TEVOC, driving school), Recon, moot court, OC spray training (Oleoresin Capsicum or better known as pepper spray).

Week 17-18...Fitness challenge week and graduation.

Hogan's Alley is a tactical training facility on the FBI Academy campus. The facility is spread over ten acres and opened in 1987. The Hogan's Alley facility was designed to provide a realistic urban setting for training agents of the FBI, the DEA and other state and local law enforcement agencies. Hogan's Alley was built as a small city with a bank, a post office, a hotel (The Dogwood Inn), a laundromat, a barber shop, a pool hall, homes, shops, a movie theater (today, the updated theater is almost an exact replica of the movie theater in Aurora, Colorado that was the site of a mass shooting on July 20, 2012 where 12 people were killed and 58 were injured during a showing of The Dark Night), and many more urban settings. The town is populated by role playing actors who are innocent bystanders, terrorists, bank robbers, drug dealers, or other criminal parts. Hogan's Alley was established after a catastrophic 1968 shootout on a Miami street where two FBI agents were killed and another five were wounded. Hogan's Alley came to be to prevent further FBI agents from the same fate. More training and realistic training was the goal of the investment to build Hogan's Alley.

Since NATs at the FBI Academy were required to always wear a

CHAPTER THIRTY-ONE

firearm during the 18-week training, Hogan's Alley was no different. Range guns were real guns, Glock 19 models with iron sites. Hogan's Alley firearms were blue guns that meant they fired ten-round magazines with Non-Lethal Training Ammunition (NLTA). NLTA bullets could break the skin but had the non-lethal capacity of a paint ball round. Paint ball hits were no picnic, but they did not kill the target. Red guns were worn on the main campus all day and were non-actionable guns. The gun range at Hogan's Alley was hallowed ground and the message to all the NATs was simply that our town protects your towns. The name Hogan's Alley was borrowed from a comic strip of the late 1890's. The alley was in a rough neighborhood, so the name fit the FBI's crime-ridden make-believe town.

Deklan Novak was the most experienced trainee on the range each day. Novak fired more rounds in SEAL training than the Hogan's Alley lead instructor, Jason Rousy ever fired as an FBI agent and as a decorated Army Ranger veteran. Special Agent Rousy was a former Army Ranger from the 75th Ranger Regiment, an elite infantry combat formation within the United States Special Operations Command (USASOC). Most Rangers had issues with the predominant assumptions from the public that Navy SEALs were the most elite military unit in the world. Ranger School was divided into three phases: Benning, Ga. Phase (Crawl Phase), Mountain Phase (Walk Phase) and Swamp Phase (Run Phase). Ranger training lasted 61 days or approximately nine weeks. SEAL training or BUD/S with Phase 1, 2, 3 and SQT lasted 65 weeks.

At Hogan's Alley, the gun range schedule was extensive, and the new agents used various ammunition loads for situational training. When facing an assailant, if the agents used non-lethal training ammunition (NLTA), helmets were required. With blank rounds, the NATs did not wear helmets. Agents were required to read the criminal actor's expressions and facial language during the confrontational practice drills. FBI agents were trained extensively to use facial expressions to understand what the criminal mind was communicating. Facial body language revealed anger, fear, surprise and deception. Hogan's Alley Gun range training was 30 four-hour sessions over ten weeks. The Hogan's Alley gun range shooting, like the initial phase shooting, took place from a firing distance to the target of 3-yards to 25-yards for pistol training, alternate hands at 3 yards (dominate and support hands), two-handed stance from 5-25 yards, standing, kneeling with and without cover. The 850 hours of training at the BFTC did not suffer fools.

On one shooting exercise late in the Hogan's Alley new agent schedule, around week 14 or 15, Deklan had taken his position behind a police vehicle and the suspect appeared holding a weapon but was at least 30 yards from the police vehicle barricade. Deklan firmed up his two-handed grip on the Glock Blue model in his hands, stood up and fired multiple rounds at the suspect with a gun clearly (or as clear as things were from 30 yards) in his hand. The instructor stopped the exercise and addressed Deklan.

"That was a bad shoot." Jason Rousy announced.

"Sir?" Deklan asked.

"Do you know for certain that the suspect had a gun in his hand?" The first question came from the lead FBI instructor at Hogan's Alley.

"Yes sir." Deklan replied. "I saw clearly the weapon was in his right hand."

"And you were comfortable firing multiple rounds at your target from in excess of 30 yards?" The second question came from Deklan's supervisor.

"Yes sir." Deklan replied again without hesitation.

"Come with me." SA Jason Rousy motioned the group to follow him to the gun range. Deklan and his fellow NATs followed as instructed.

The group reached the gun range and SA Rousy retrieved one loaded real Glock from the gun room with a couple full magazines. The group settled in behind the shooting stalls. They knew what was coming.

"Agent Novak, step up to the mike." Deklan's instructor ordered, using a microphone reference sarcastically. "Send down the target to 30 yards. You fired at our target outside from approximately the same distance. Shots missed are shots that can kill innocent bystanders. I

CHAPTER THIRTY-ONE

know you are a star on the range under no stress conditions. You may have been confident in your eyes to be certain the man had a gun and was a threat, but you were still too far away to use deadly force. Let's see how many shots find the torso on this target from 30 yards."

Special Agent Jason Rousy backed away and crossed his arms like Gunnery Sergeant Hartman from **Full Metal Jacket**. Deklan stepped up to the counter where the Glock and the magazines had been placed. The live-round Glock magazines held 17 shots. Deklan calmly loaded the first mag into the pistol and took his stance. Deklan Novak emptied the weapon at the target. He fired in succession but not frantically. The pistol barely moved as the explosions sounded like a choreographed audio dance sequence. The gun barrel froze after the final round left the chamber. Deklan put the gun down after he pulled out the empty magazine. Deklan pressed the return button for the target and grabbed the stiff paper target with a man's torso outlined as the bad guy. Novak handed the target to his instructor, who examined the results. Deklan had fired 17 rounds at the target. There were 17 rounds within the inner section of the outlined torso. Deklan had basically fired 17 bull's eyes from 30 yards away.

"We like to reference the 21-foot-rule in training for an assailant that is coming at you." The FBI instructor stated after viewing the impressive target results. "If an assailant pulls a knife while running at you and he is 20 feet from you, chances are good that you have time to draw your weapon and fire before you get stabbed. That is basically crap. Close encounters end poorly. Accuracy research shows that officers hit their targets 14% of the time from 2-10 feet away and hit their targets 65% of the time from 21-32 feet away. NAT Novak is the outlier among us, but while Novak can shoot the eyes out of a spider from a football field away, he will be crucified in a judicial avalanche if he ever fires a weapon in the line of duty from that distance and misses the bad guy. Great target shooting today, Deklan, but when I tell you that your shoot was bad, I am not asking for a debate." The message was delivered loud and clear. Jason Rousey was the real deal.

The team practiced situational shooting for weeks. What was a good shooting? What demanded or called for deadly force? Where was an agent justified in using deadly force and where was the agent prohibited from using deadly force? The situational drills were essential to the actionable real-life threats agents faced every day. Their reactions had to be legal first and foremost. If an assailant was running away, can the agent fire at the man and shoot him in the back? Yes, if the assailant had a weapon and was searching to gain a tactical advantageous position to attack innocent bystanders or any law enforcement people on the scene. Can an agent shoot an unarmed assailant? Yes, if the assailant was known to have skills that threatened the life of the agent or threatened the lives of the law enforcement agents on the scene. FBI agents that found themselves facing imminent danger of death or serious physical injury were justified in the use of deadly force. For example: If an FBI agent was trying to arrest a UFC champion fighter like Chuck Liddell or a heavyweight champion like Mike Tyson, those individuals possessed skills that presented a clear and present danger to the agent. If the assailant moved towards the agent, legally the agent was allowed to put the assailant down.

All the decisions taught at Hogan's Alley and used in the field were eventually subjected to an FBI review board for certain and most likely would be subjected to a courtroom and a judge. The legal use of deadly force in our society is based on the subjective interpretation of the laws as they were written. If Deklan was in the field working as an FBI agent, where he fired at a suspect from 30 yards away and shot an innocent bystander from that distance, there would be a certain tsunami of accusations, reviews, charges and lawsuits to follow. The attorneys representing the victim's family will proceed with a civil rights lawsuit, a wrongful death lawsuit, an excessive force lawsuit, a negligence lawsuit or any combination of the suits mentioned. In addition to the lawsuits filed by the victim's family, possible reckless homicide or homicide charges could be filed against Deklan by the DA. Charges are determined by the local district attorney's office which is often politically influenced by public outrage. A frequent example is when a young man is shot by police under the claim of imminent danger, the young man's eighth grade graduation photo is shown on television stations repeatedly. The image of an innocent

CHAPTER THIRTY-ONE

young boy massacred by law enforcement is the prevailing sentiment presented. No mug shot photos from the 17 times the suspect was arrested are ever shown on television. Officers and agents are often charged with murder.

Deklan stood outside and the weather was not cooperating. The wind was howling, and the cold air cut through the skin like a serrated knife. It was Novak's turn to get blasted in the face with oleoresin capsicum or pepper spray shampoo. The instructor asked if Deklan was ready, and Novak was certain the instructor let loose before he nodded. Deklan's face was on fire, but Deklan was required to open at least one eye, attack a heavy bag and defend himself against an assailant who was trying to wrestle the gun from his holster. Deklan overpowered the spray and the assailant. The test was not comfortable, but Novak passed easily. The pepper spray exercise was one of the final tests at the Academy. Nearly 1,000 hours had been spent in classrooms, on gun ranges, on self-defense mats, in vehicles hurtling out of control, in movie theaters, at front doors attempting to arrest an accused felon, inside a bank lobby negotiating with a robbery suspect, testifying inside a mock courtroom, and then training like a madman on his own time after dinner to keep his body in tune with his chosen profession. The class had come closer together as they were humiliated, praised, enlightened and trained as a team. They finally would stand together and receive their FBI badges and credentials together.

The graduation hall held 996 guests and the place was packed. Deklan's graduating class had 202 graduates. Deklan was a part of Charlie section. Alpha, Bravo and Delta joined the festivities. Deklan's mother, Julia, could not bring the boys, Jack, James, Hank and Tony. Each graduate got four tickets, and the numbers did not work out for Deklan and his sons. Deklan's sister, Dana, came with Julia. The other two tickets went to Behr Thomas and Charlie "Cobra" Coletti, Deklan's Hell Week tormentor and eventual best friend. The FBI Director Mark Beatty and the FBI Assistant Director Tania Crawford were on the podium. Assistant Director Crawford would emcee the ceremony, and the FBI Director would speak at length. Tania Crawford was the first female to achieve the position of Assistant Director of the entire agency. Birdie Pasenelli became the first female assistant director at Headquarters, in charge of the Finance Division. In 2001, Kathleen McChesney attained the rank of executive assistant director. Assistant Director in Charge of the Chicago field office Monica Corcoran also boosted the female ascension ranks.

The 90-minute ceremony was reminiscent of when Deklan received his Trident. The length of training time was not close. The physical challenges were not close. When Deklan held his gold FBI badge and his FBI credentials in his hands, the exhilaration was damn close. Staring at the gold FBI badge and his identification credentials, Deklan filled his lungs with air and exhaled slowly. People often asked Deklan what he was trying to prove along his improbable journey. Deklan smiled and thought to himself…" not a fucking thing that you could understand."

The final stop the new FBI Special Agents was at the weapons vault to pick up their FBI issued Glock 19, and the ammunition for their new firearm. Deklan and his mother and sister took pictures with FBI Director Mark Beatty. Coletti refused to join the photo op. No surprise to Deklan as he flipped Coletti the bird. Behr Thomas and Deklan embraced and moved ahead without many words. Behr Thomas hid the few tears he shed during the ceremony. Thomas had his daughter back living in his home because Deklan Novak had entered his life.

CHAPTER THIRTY-TWO

2018...ASSIGNMENT TO THE FBI CHICAGO FIELD OFFICE

THE CHICAGO FBI field office was in the Bridgeport neighborhood in Chicago. Like Boston's Beacon Hill is synonymous with the city, Bridgeport is the historical reference when anyone mentions the long line of Chicago's mayors. Bridgeport has been the home or birthplace for five Chicago mayors representing all but 10 years between 1933 and 2011. The mayors that hailed from Bridgeport were Edward Kelly, Martin Kennelly, Richard J. Daley, Michael Bilandic, and Richard M. Daley. The FBI was the sole tenant inside the ten-story white office building that curved in a half-moon arc on the manicured grounds. The property was well groomed with a black wrought iron, eight-foot-tall fence surrounding the five-acre space. There was one drive-through entrance at the right end of the circular drive that allowed visitors to get their credentials in order before approaching the guard house.

The Federal Bureau of Investigation operated 56 field offices in major cities across the United States. Many of these offices were further divided into smaller resident agencies that had jurisdiction over a specific area. The hierarchy at the Chicago FBI field office was one of the larger units in the country. Chicago, New York City, Washington D.C., and Los Angeles were managed by an Assistant Director in Charge (ADIC), followed by the Special Agent in Charge (SAC or SAIC), then the Assistant Special Agent in Charge (ASAC), Supervisory Special Agent in Charge, Senior Special Agent, Special Agents and Special Agents from the Academy on probation. The Chicago FBI field office was home to 377 people, including the command staff, special agents, intelligence analysts, support professionals and clerical staff members.

Deklan Novak arrived at the Chicago FBI field office one week after his Academy graduation. Deklan carried the Glock 19 handed out in Quantico. The former Navy SEAL and newly assigned FBI Special Agent requested an M-4 rifle as soon as he settled in. Deklan was not nervous. SA Novak was anxious to meet his VCAC squad. Deklan was dressed professionally and conservatively but the FBI suit and tie had become a retired requisite look. Suits were worn predominantly in court, but street agents were instructed to blend in with the population.

Deklan Novak had a unique assignment as he was tagged to go directly to the Violent Crimes Against Children (VCAC) squad in Chicago, a tactical position born from the Violent Crimes Apprehension Program (VICAP) within the FBI's National Center for the Analysis of Violent Behavior (NCAVC). The mission of the VICAP is to assist law enforcement around the world to identify and apprehend violent criminals. Deklan would have a training agent veteran with at least five years on the job as his mentor. New agents, regardless of their fast-tracked assignments, were required to work with a veteran agent for at least one year. The veteran agent chosen to work with Deklan was younger than Deklan. Deklan's senior agent mentor was SA Brett Moore, 30 years-old and from Austin, Texas. Brett Moore was sent to the Chicago office as his first assignment after graduating from the academy. The mentor's job, among other things, was to teach the new agent how the FBI operated, from paperwork to booking a criminal. Major city FBI field offices had their own holding cells. Brett Moore was also a member of the Chicago based VCAC squad.

The assignment to the VCAC squad right out of the academy was

CHAPTER THIRTY-TWO

considered a plum assignment to say the least, but the assignment did not preclude Deklan Novak from receiving the new kid on the block duties or the grunt work at the field office. The Special Operations Group (SOG) was much more impressive sounding than the daily requirements assigned to the title. Deklan sat on the desk overnight to monitor online complaints. Deklan was given every call-in complaint that had to be checked out by a special agent. Calls that needed an agent visit included internet financial extortion of a minor. The financial extortion of a minor had grown exponentially over the past ten years. The crimes began when a criminal posing as a minor reached out on social media to a minor female high school or junior high school student and initiated a "girl/girl" relationship. Over the course of that relationship, the criminal obtained compromising photos of the minor and then threatened the minor with sending the photos viral unless a ransom was paid to hold the photos.

Other scams that required a Deklan Novak visit included many elderly scams where criminals disguised a voice through AI and posed as a grandchild or relative's voice on the phone call. The caller, impersonating the grandchild or relative, claimed to be in trouble…either from an abduction or they had landed in jail. These scams were called virtual scams. The caller sought immediate payment to get out of jail as bond or to be released from the imaginary abductors. Other more complicated scams claimed a relative had been involved in a car accident. The caller claimed the relative hit his 17-year-old son on a motorcycle and the grandson or granddaughter needed to pay $3,000 in ransom for the damage or the caller would put a bullet in the relative's head. Most elderly people receiving a call like that were too confused and scared to doubt the story. AI voiced disguises provided all the necessary credibility for scared grandparents. Scams directed towards the elderly preyed on their kindness, family loyalty and vulnerability. Numbers that popped up as local numbers were often from overseas.

There were many bomb threats called into a major city FBI field office. Most big city bomb target threats were directed at schools, universities, government buildings, malls, private businesses, public figures and political or ideological targets such as Planned Parenthood. Deklan was sent to investigate the validity of the bomb threat calls. The standard procedure model for the FBI regarding bomb threats was to take the calls seriously although the statistics revealed to law enforcement agencies that the people making bomb threats were largely separate from those who attempted a real bombing. Real bombings typically occurred without warning.

During Deklan's first week at the Chicago field office, a coked out-of-her-mind assailant had made her way past the guard house and tried to break in through the doors to the FBI offices. Deklan was manning the main desk computer after hours, trolling networks like Tor for child pornography sites and sat up suddenly. The office was closed, and the attempted break-in was a shocking interruption to what had been a quiet night.

"What the fuck?" The former SEAL and newly assigned FBI Special Agent jumped out of his chair and approached the commotion at the main entrance.

The woman did not show a weapon, so Deklan kept his Glock holstered although he had one hand on the pistol. Deklan yelled for the woman to back-up and to move away from the door. The female complied and backed up, temporarily pausing her breach of the FBI's front door. Deklan opened the door and continued his motioning for the female to remain where she was. The female assailant had other ideas. The woman charged Deklan and attacked the new Special Agent like he was back in an MMA octagon. The assailant was screaming…" fuck the FBI," as she tried to reach for Deklan's firearm while kicking and screaming mostly incoherent references to some higher deity. Deklan was more flabbergasted than threatened but had to end the confrontation quickly. Special Agent Deklan Novak backed up one step from the female Tasmanian maniac attempting to assault a federal officer. Deklan prepared for another frontal assault and as the woman came forward, Novak stepped quickly around her and turned the light female frame easily, so her back was next to his chest. Deklan quickly slipped his right arm around the woman's neck and his left hand pressed hard against the back of her head completing a

CHAPTER THIRTY-TWO

perfect rear-naked choke hold. Deklan applied hard pressure for maybe ten seconds. The meth-head or coked-out woman dropped to the lobby floor instantly. The female suspect was unconscious and appeared to be sound asleep. Deklan looked down at the motionless figure on the ground.

"What the fuck?" he repeated. So much for the Director's personal appointment to the Chicago field office and the coveted VCAC squad.

The Chicago VCAC squad consisted of Supervisory Special Agent (SSA) Jesse Goodman, a ten-year FBI veteran and former Hostage Rescue Team (HRT) member. Brett Moore, Glenn Bradley, Kristin Mullins, Christos Giannaras and Linda Kingsbury. The squad awaited the new member with skepticism and a fair level of apprehension. The VCAC squad members were unmarried, and they were not looking for a rock star former Navy SEAL or a parent with four boys to join their ranks. The horror witnessed daily as a member of the squad that investigated violent crimes against children was not a good image to bring home to the wife and kids.

During Novak's first week in Chicago, the VCAC squad conducted an arrest at a high-rise apartment building in River North. The man they sought with a federal warrant was Arturo Perez, from Amarillo, Texas originally. According to the warrant, Perez was distributing videos on an invitation only forum on the dark web. The videos showed multiple males sexually abusing children three to four years old. The complaint detailed the graphic violence in which Perez and two other males present sexually abused two children between three and four years old. The National Center on Child Abuse and Neglect (NCCAN) shared the videos with the FBI as they contained two common identifiers: two gold bracelets and a pair of white Converse shoes. The FBI was able to match one of the toddlers to a relative who posted photos of the missing child on social media. Perez had worked at the home of one child in the video on more than one occasion but had not been a suspect when the child went missing. A search through Arturo Perez's Instagram account revealed a reference to the home of the second child in the video. The FBI taught Deklan and all their agents to never underestimate the stupidity of internet criminals. In other words, never ignore in an investigation something that, on the surface, appears obvious, like posting a photo of a vintage Mustang from the second missing child's residence on Arturo Perez's Instagram account.

Deklan Novak was thrown into the fire right away. The squad needed to know who the new guy was. The Chicago VCAC squad had taken down more than 30 people over a three-year span following one of the largest dark web child porn marketplaces that spanned more than 28 countries. Deklan's extensive experience as a breacher gave him no jitters about leading the tactical response team to the Perez apartment and executing a dangerous warrant. Jesse Goodman led his VCAC squad into the exquisite lobby, followed by Deklan Novak and Brett Moore.

The Perez luxury apartment building, The Parthenon, at 169 W. Huron was two blocks from the Rock and Roll McDonalds, the Hard Rock Café on Ontario Street and the Excalibur Nightclub at the corner of Dearborn and Ontario in River North. The massive nightclub, Excalibur, opened in 1989 and was located inside a Chicago landmark building that was the original Chicago History Museum, then became home to the Chicago Historical Society and finally became the Chicago location for the Limelight Nightclub which opened in 1985. The Limelight nightclub originated in Florida in the 70's, then opened in New York City in 1983 and was owned by Peter Gatien. The Chicago behemoth nightclub was three levels and 60,000 square feet. The River North neighborhood in Chicago was trendy, a destination spot for visitors to the city and extremely upscale. The Arturo Perez apartment building's lobby looked like the Taj Mahal, an ivory-white marble mausoleum. The lobby ceilings were 18 feet tall, and the floor, walls and front desk were all constructed from what appeared to be various shades of white marble or granite.

The doorman and the front desk manager backed away as the FBI badges were shown and the team proceeded to the elevators. Perez lived on the tenth floor of the fourteen-floor building. The building had a rooftop lap pool, a large community health club and lounge with gaming tables, an outdoor kitchen with grill stations, a pet run

CHAPTER THIRTY-TWO

and pet wash and a yoga studio. The building was about to have a vacancy. The team gathered outside the elevators and around the corner from the door to the Perez apartment. Jesse Goodman checked his Glock 19 with a Trijicon HD XR Night Red Dot Laser Sight. Deklan carried the same Laser Sight set-up on his Glock 19. FBI special agents normally did not pull their guns until they were stacked outside the door they were about to breach. Once stacked at the door, guns were held in the SUL position, an extended ready position that offered good weapon retention, good situational awareness and kept the gun pointed in a safe position and downward angle while in a high ready state.

"Ever since I was a kid, I wanted to be in the FBI." Jesse Goodman said under his breath to Deklan Novak.

"What?" Deklan heard his boss but wasn't quite sure how to respond. Novak pointed his laser sight at the wall across the hall and watched it dance on que from the Frogman's steady hand. Deklan holstered his weapon quickly. Deklan grabbed the 30-pound Pocket Ram with a 5" by 5" strike plate.

"You heard me." SA Goodman replied. "I wanted to be in the FBI to bring down shit-bags like this guy. The scumbag here hosted an internet site with 215,000 users in 28 countries, posted 23,000 sexually explicit images and videos with toddlers and was making enough money from that to live at this address. At least a half-million laptops, tablets, and phones are viewing or sharing child pornography on the internet every month." Goodman was ready to approach the apartment. "We will operate professionally at all times, but where is your line, Novak?"

"I'm on my first arrest with my new field office squad." Novak mentioned. "My line is what you tell me my line is."

"Fair." Goodman countered. "What was your line overseas in combat?"

"I wouldn't lose any sleep if Perez got put down." Deklan answered while thinking immediately of the one man he had recently shot.

"This is a ten-grand a month building minimum…for a one-bedroom. A hundred bucks says he's got a car in the garage that costs more than I make in two years." Goodman noticed the rest of his team was listening. Brett Moore and Glenn Bradley were closest to Goodman.

"We can't tell the new guy that we are hoping a suspect pulls a gun on us, boss." Bradley mentioned while the banter was hardly casual. The FBI veteran team members knew that men like Arturo Perez employed the best legal team that money could buy.

"We can't?" Brett Moore asked with a tinge of disappointment in his voice. Goodman winked and held his weapon down along his hip.

"Johnny Cochran charged O.J $675.00 per hour. Arturo Perez was arrested in 2020 for selling a video of him ejaculating on a seven-year-old girl. The case was thrown out of court because his lawyers claimed his computer was inappropriately accessed. Let's go get the pedophile asshole." Goodman flipped his head forward to indicate the group was moving. Deklan Novak lined up first as he was the breacher. Goodman followed. "None of us know if there are children inside the premises. If we announce that we are police officers and ask for permission to enter the apartment, we may put children at risk. Novak, get ready." Goodman spoke to Novak and the squad. Deklan turned back and nodded. The VCAC squad positioned themselves in a stacked lineup.

"I'm one." Goodman said as he held up one finger and was poised to enter the apartment first after the door was breached.

"I'm two." SA Moore said softly and held up two fingers. Weapons were ready. Red laser dots scattered along the floor outside the Perez apartment.

"I hope he does." SA Goodman said a moment later as he moved forward towards the door.

"Does what?" SA Novak asked quietly.

"Resist." The squad leader uttered as he reached the front door to the Perez apartment.

The squad had rehearsed the "forced entry" prior to their arrival in River North. Based on the nature of the warrant, where children were in danger, the VCAC squad did not allow the suspect to hurt a child as a potential witness or to destroy any evidence by announcing their

CHAPTER THIRTY-TWO

arrival as the FBI. Goodman's VCAC squad did not ask for permission to enter the premises. Novak was very well-versed in using a "Pocket Ram" breaching tool. The Pocket Ram defeated the locking mechanism on the front door with ease. The front door to the Perez apartment blew open like it had been lined with C-4. Deklan, Moore and Goodman took down the suspect in less than 30 seconds. SA Goodman had no more doubts about the new squad member. Arturo Perez resisted arrest and sustained some minor cuts, two broken ribs and a shattered nose. An excessive force complaint was filed against SA Jesse Goodman and the VCAC squad with the federal Inspector General's office. After a brief investigation, the complaint was thrown out. Arturo Perez was tried and convicted of child sexual abuse, child abduction, producing and distributing child pornography and the obscene visual representations of the sexual abuse of a child. Arturo Perez was sentenced to life in prison.

CHAPTER THIRTY-THREE

2024...A CHILD ABDUCTION AT THE STARLIGHT SKY ACADEMY

IN 1995, the Starlight Sky Academy began operating as a summer daycare center. In 1995, the school was a small, summer inexpensive childcare option for young parents in a sea of larger, more prestigious daycare centers. The Starlight Sky Academy opened in a vacant bicycle shop, with no more than 2,000 square feet of barren space, near the train station on Linden Street in Wilmette, Illinois. From that small summer daycare center, Starlight Sky grew over the years and became a year-round daycare business that operated as the current Starlight Sky Academy.

The school boasted eleven classrooms on three floors and an enrollment of nearly 200 children, spanning 6 weeks old to 12 years old or the fourth grade. Tuition at the Wilmette, Illinois daycare center was $680 per week for infants 6 weeks old to two years old. Tuition for three-year-olds and up was $460 per week. The center opened at 6:30 a.m. and closed at 6:00 p.m., Monday through Friday. Parents were able to drop off the children and pick them up at any time between the hours of 6:30 a.m. and 6:00 p.m.

Security at the Starlight Sky Academy was diligent, but vulnerable at the same time. There was one main entrance at the front of the school that was the security checkpoint to enter the building. The first door opened to a small lobby manned always by a school official behind a security glass window. The parent or visitor was required to be checked into the school from an appropriate list. If the visitor was not on a preapproved list of visitors, the visitor was not allowed to enter the school. Parents were required per DCFS and the academy rules to fill out carpool forms and authorization forms detailing anyone allowed to pick up the child. Social Security numbers and driver's license numbers were required from each authorized carpool member. Each classroom and each teacher had a copy of the authorized carpool list. The employees had key fobs to enter each floor of the school, administrative offices, supply rooms, snack rooms, the kitchen, a nursing room and every classroom. Restrooms were not locked. There were 18 cameras on the grounds covering each classroom, the hallways, the outside grounds front and back, and the outdoor playground. The playground was a small enclosed outdoor area with only one way in or out. The one playground entrance was only accessible from the back classroom on the main floor, Room #110. The security monitor double-screen display inside the main office had 18 sections showing each camera angle. The security camera monitor was not manned 24/7, although the cameras operated 24/7 and the tapes reset on a weekly basis.

The Starlight Sky Academy did not employ a security guard. The issue was a constant subject of debate at the monthly schoolboard meetings. There have been eleven mass school shootings nationwide since Columbine, Colorado in 1999. The images of students running from their school with their hands up has become a haunting familiar sight on the nightly news. The academy had a working relationship with the Wilmette Police Department. There was a Wilmette PD officer present each weekday morning for the drop-off time frame between 6:30 a.m. and 8:30 a.m. Pick-up times varied too much after lunch to have an officer present for every parental pickup time during the afternoon. Wilmette PD had an officer at the school from 3 p.m. until 4 p.m. each school day.

Of the 150 or so families with children at the Academy, one prom-

CHAPTER THIRTY-THREE

inent member with two grandchildren at the Starlight Sky Academy was Bryce Stanley, founder of the Stanley Law Group, leaders in personal injury and wrongful death litigation with offices based in Chicago. The Stanley Law Group had over 40 years of experience representing aviation accident victims and their family members, as well as victims of medical malpractice and personal injury accidents. Offices for the Stanley Law group were in Two Prudential Plaza, a 64-story skyscraper on Lake Street in the Loop area of Chicago, Illinois. Bryce Stanley owned a palatial home on Sheridan Road in Winnetka, Illinois.

June 2000

New Trier East High School in Winnetka, Illinois was holding their Graduation ceremony for the millennial year 2000 at Northwestern University's Welsh-Ryan Arena as they had done for many years. The graduating classes at New Trier had gotten too big to hold the graduation ceremonies on the Winnetka campus or the sister campus, New Trier West, in Northfield, Ill. Welsh-Ryan Arena had a seating capacity of 8,117, more than enough room to accommodate the 1357 graduating seniors and their families. Luke Rice, Andrew Stanley and Grayson Maclagan grew up together in Winnetka and were graduating high school together on a smoldering hot and humid Saturday in early June 2000. Andrew Stanley was Deklan Novak's half-brother and the subject of a 2014 Chicago Daily-Times feature by reporter Xander Moss.

After a wild night of parties with their friends, the three boys met at the Stanley home in the morning. The families of each boy were gathered at the Stanley home for a proper send-off. The parents sipped coffee and talked about how fast the years had flown by. The boys inhaled the sweet rolls and doughnuts spread out on the kitchen table. Siblings and a few assorted grandparents in town for the graduation festivities all told the boys to have fun but not too much fun. Kisses and hugs were exchanged. Golf clubs and duffle bags were packed into the Range Rover. Andrew checked the gas tank. The boys left on the morning of June 7, 2000. They headed down Willow Road to the Edens Expressway. The route took them to the Kennedy

Expressway dissecting downtown Chicago. The morning was still hot and humid, unusual for early June in Chicago. The Range Rover departed the driveway just after 10:00 a.m. Andrew was driving. Luke rode shotgun in the front seat. Grayson had his headphones on in the back seat. Andrew's father, Bryce Stanley, wondered if allowing the boys to take a $90,000 vehicle was such a great idea. He smiled as everyone waved good-bye. The destination was a Myrtle Beach golf trip. The boys were never seen again. Three young men vanished from the face of the earth. There was a security video that showed the boys at a gas station in Lexington, Kentucky on the afternoon of June 7, 2000. The boys have been missing for more than 20 years.

Present day...2024

Bryce Stanley had one surviving son, Logan. Logan and his wife, Hillary had two children attending the Starlight Sky Academy. Luca was two years old, and Layla was four years old. The call came in just before 11:00 a.m. on a Thursday morning. The morning session was set to end at 11:45 a.m. Hillary Stanley was an investment broker for Merrill in Winnetka, Illinois. Hillary's father-in-law, Bryce Stanley was one of the wealthiest lawyers in Chicago. As the managing partner in one of the most successful law firms in Chicago, the Stanley family was rich and a North Shore fixture among the charity ball circuit. The family contributed generously to the Wish Ball for the Make-A-Wish Foundation, the Goodman Theater Gala, the Lincoln Park Zoo Ball, the Ravinia Gala, and the American Cancer Society Skyline Soiree. The two-year old student daycare program at Starlight ended at 11:30 a.m. and the Jr. K program for four-year-olds ended at 11:45 a.m. The kids waited inside a supervised enclosed walkway where parents parked and came up to the school to pick up their children. The Starlight employee supervisors released the children to the approved parent.

The call from Hillary Stanley told the front desk at the Starlight Sky Academy that she was going to be unable to pick up Luca and Layla at 11:45 a.m. Hillary explained that her meeting downtown ran very late and her neighbor, Brittany Hawn was going to pick up the children. The caller ID number was Hillary Stanley's phone number.

CHAPTER THIRTY-THREE

The front desk attendant noted the call in the logbook. Brittany Hawn was on the list as an authorized carpool member.

"I'm so sorry." Hillary Sb x3tanley explained apologetically. "I was positive my meeting was going to end by 10:00 a.m."

"That's fine, Mrs. Stanley." The front desk attendant replied. "Thank you for calling. We will watch for Ms. Hawn to come for the children."

"Thank you, again." Mrs. Stanley repeated. "Good-bye."

"Good-bye, Mrs. Stanley."

At 11:40 a.m., a white 2014 CL-Class Mercedes Benz, 2-door coupe, valued at $116,000 pulled up to the Starlight Sky Academy to pick up Luca and Layla Stanley. The car was also on the authorized carpool list under the Stanley family. Brittany Hawn got out of the car and walked quickly to the front entrance of the center and where the children were waiting.

"Hi, I'm Brittany Hawn, Hillary Stanley's neighbor here to pick up Luca and Layla." Ms. Hawn explained to the school attendant working in the waiting area.

"Hello, Ms. Hawn." The attendant replied. "Can I see your driver's license?"

"Certainly." Brittany Hawn pulled out the license. The attendant glanced at the license and then back to the blonde lady with a ponytail and aviators. "I left in a hurry after Hillary called. She told me to use the white Mercedes because the car was well-known at the school. I jumped in Hillary's car. I live at 990 Sheridan Road. We are next door to the Stanleys."

"Wait here, please." The attendant walked to the glass window by the main entrance and spoke to the woman inside the office. The addresses checked out. The car checked out and the woman was also on the authorized list. The video security cameras recorded all pickup activity.

"Sorry for the delay, Ms. Hawn." The attendant explained. "We must check our records. Thank you for your patience and have a great day." The attendant got the two children and brought them to Hillary's neighbor. The young kids looked at their neighbor and Ms.

Hawn gave them each a candy bar. Chocolate was the great equalizer. The Starlight staff headed back to the school.

"Hi." Brittany Hawn said to the kids softly. "I'm from your mom's office." Luca and Layla saw their mother's car at the curb.

Brittany Hawn picked up Luca and grabbed Layla's hand. The group walked back to the Mercedes at the curb. Ms. Hawn hooked Luca up into the car seat in the back seat of the car and Layla was belted into the front seat while devouring the chocolate bar that she was given before they arrived at the car. Luca had already tested the chocolate bar as a make-up utensil. The expensive German vehicle pulled away from the curb and headed for Ridge Avenue.

Deklan Novak had rented an apartment in the West Loop after he returned to Chicago from Quantico. The apartment was at 728 W. Jackson, not far from the United Center in Chicago, where the Chicago Bulls and the Chicago Blackhawks played their home games. Novak was in no hurry to buy a place and renting worked for his first FBI assignment. Deklan renewed his lease term each time it expired. Deklan's apartment was a two-bedroom, two bath, 1653 square foot apartment on the 12th floor and Deklan paid $4900.00 per month for the apartment. Deklan had a decent net worth, but as a first year FBI Special Agent, he made a base salary in Chicago of $94,000 annually. The cost of living in Chicago was high and the FBI base salaries reflected the Consumer Price Index (CPI), at least minimally, for the location of the FBI new agent assignment. Special Agents in Los Angeles, Chicago or NYC made more money than new Special Agents in Sioux Falls, South Dakota. When an agent agreed to work a minimum of 50 hours per week, the agent's salary jumped 25%. (FBI agents were not paid by the hour, but the bump in pay was the FBI equivalent to overtime pay.) Deklan was making $117,500 per year with the 25% bump. No one was getting rich as an FBI Special Agent, but no one was walking a bread line, either.

Deklan was off on Thursday and Sunday for the current rotation they had been working on at the Chicago FBI field office. The hours always came to more than 50 hours per week, but Novak loved every day on the job, regardless of the unscripted, call-in assignments he

CHAPTER THIRTY-THREE

was given as the new guy. Deklan had been to the Cross-Town Fitness Center for a morning workout. He had showered and was about to grab some lunch on Madison Street when the security message call came in for Deklan to immediately head to work. Deklan arrived at the Chicago field office at 1:30 p.m. The VCAC squad was assembled. Jesse Goodman addressed the squad.

"There were two children abducted this morning at 11:45 a.m. from the Starlight Sky Academy in Wilmette. The children's mother called in at 11:00 a.m. to let the school know that her neighbor was going to pick up the kids today because her meeting was running late downtown. The mother is an investment broker for Merrill in Winnetka. Her name is Hillary Stanley. Her father-in-law is Bryce Stanley." Jesse Goodman paused because everyone knew who Bryce Stanley was.

"The caller ID was Hillary's phone number." Jesse Goodman continued. "The neighbor was on the authorized list and was a look-alike for the neighbor. The only issue was that one of the school employees at the main pickup area thought something looked unusual when the kids got in the car with the neighbor. The school employee thought the older child tried to get out of the car as they pulled away. She said the neighbor appeared to grab the child roughly and sped away. The school supervisor called Wilmette PD. They notified our office. No one can locate or reach Hillary Stanley or the neighbor, Brittany Hawn. Logan, Hillary's husband and Bryce Stanley were notified and they are on the way to the Starlight Sky Academy."

"Boss." Deklan spoke up.

"Yes, Special Agent Novak." Jesse Goodman responded.

"The Starlight Sky Academy is owned by my mother." Deklan revealed.

"I know, Special Agent Novak." Goodman replied. "The Starlight Sky Academy is owned by Starlight, LLC. The managing member and the Director of the Starlight Sky Academy is Julia Novak. Agent Novak, please let us know anything relevant to the missing two children that you can contribute to the upcoming investigation. Andrew Stanly, Bryce Stanley's stepson, has been missing for two decades.

Andrew Stanley was your half-brother. Please explain Special Agent Novak, why these facts are or are not relevant." Jesse Goodman raised his eyebrows and stayed silent.

"Andrew Stanley and two friends left Winnetka after they graduated from New Trier High School in 2000 for a golf trip to Myrtle Beach." Deklan replied to SA Jesse Goodman and the VCAC squad present. "The three boys were never seen or heard from again. The FBI was very involved in the subsequent search. Andrew Stanley was my half-brother. What this has to do with today's kidnapping and my mother's school is perhaps a coincidence, but more likely is related to another case that my father was involved with in the nineties. The case was handled by the Stanley Law Group." Novak recalled the 20-year-old case.

"My father's sister, my Aunt Paige Novak Greene, was 47 years old and listed on an organ-donor waiting list, waiting for a kidney transplant in 1995. The average wait was five years to find a kidney suitable and available for a transplant. Some patients never received an organ. A woman roughly the same age as my Aunt Paige, Bianca Amato, was placed on the kidney transplant waiting list in 1998. Bianca Amato waited three months and was scheduled for transplant surgery, leap-frogging Paige Novak Greene and all the other patients waiting for kidneys within the North Suburban University Health System. My father contacted Bryce Stanley. They had history. The Stanley Law Group filed an immediate lawsuit against the United Network for Organ Transplants (UNOT) and the North Suburban University Health System (NSUHS). A temporary restraining order was issued in federal court that stopped the Bianca Amato surgery. Bianca Amato passed away during the lawsuit. Paige Novak Greene also passed away before the case ended. What no one knew was that Bianca Amato was Sam Giancana's niece. Sam Giancana, as we all know, was the notorious Chicago mobster and boss of the Chicago Outfit from 1957 to 1966. Giancana was killed on June 19, 1975, shortly before he was to appear before the Church Committee which was investigating CIA and Cosa Nostra collusion." Deklan stopped to gather more thoughts from what happened more than two decades past.

"Wow." SA Goodman commented. "You have my undivided attention."

"Apparently, Bianca Amato retained some serious clout within the Outfit after her uncle was killed. Joseph "the Clown" Lombardo was released from prison in 1992, but still had Bianca Amato's best interests under his control. The United Network for Organ Transplants (UNOT), the non-profit that runs the nation's organ distribution network claimed that Bianca Amato was days away from dying and therefore was able to "jump the line" and placed on a surgical schedule. Joseph Lombardo's involvement was never proven. The move up the transplant list angered my father, and he contacted the Stanley Law Group to represent Paige Novak Greene and sue UNOT. Stanley's law firm was able to block the transplant and subsequently, Bianca Amato died shortly after the legal battle began. The Outfit was not happy about the lawsuit. Many investigative theories floated around after the three boys disappeared on a golf trip in 2000. One theory had the Outfit killing the boys to get even for Bianca's death. Andrew Stanley was Bryce Stanley's stepson and my father's biological son. My father initiated the lawsuit. Bryce Stanley took the case. When the Outfit was compromised or targeted, retribution was assured." Deklan explained the possible continued connection.

"How does the abduction today figure into the Outfit theory?" SA Goodman asked and had a good idea what the answer was going to be.

"One thing is certain when the Outfit targets an enemy." Deklan remembered the case and the investigations that followed. "When you dishonor the Outfit intentionally or unintentionally, as Bryce Stanley and my father did, the Outfit will never leave your life. They may pause for a decade or two, but make no mistake, they have not walked away. The most likely method to survive the Outfit's long reach was to wait them out. Unlike the Capone era and Giancana's run, the Outfit bosses today, are not old-school wise guys. The Mob has fewer foot soldiers and street crews. Legalized gambling and legalized marijuana caused a shake-up inside the Outfit. Death and more profitable ventures surrounding local casinos were the best medicine for a cessation of the actions directed at you and yours." Deklan revealed a realistic knowledge of organized crime in Chicago.

"Do you think the Stanley children were abducted by the Chicago Outfit?" Jesse Goodman asked.

"I believe that there is a distinct possibility the Outfit did not walk away from Bryce Stanley." Deklan offered. "Mr. Stanley has a high profile in Chicago. He and his wife are staples online and in the newspapers. Logan and Hillary Stanley are prominent socialites on the North Shore. The Stanleys are photographed at every lavish charity event in the city. In addition to Stanley's high visibility, Bryce Stanley made a fortune from the Greene/Amato lawsuit."

Bryce Stanley arrived at the Starlight Sky Academy in the early afternoon. He arrived alone after the drive from his downtown law offices. Logan had gone to his home first and was on his way to the Starlight Academy. The FBI VCAC squad was on hand at Starlight. Deklan's mother, the Starlight Sky Academy Director, Julia Novak was standing next to her son. The Wilmette Police Department had numerous police vehicles parked in front of the school. Wilmette was an affluent enclave north of Chicago. The police car activity was noticed by more than a few residents living near the school. The streets were lined by century old trees and the driveways led to seven figure homes covered with ivy and sporting third or fourth floor dormer windows. Jesse Goodman explained to Bryce Stanley the circumstances surrounding the abduction of Bryce's grandchildren. Hillary Stanley was missing and so was Brittany Hawn. SA Goodman was about to begin his questioning of Bryce Stanley.

"My condolences on the loss of your son back in 2000. Special Agent Deklan Novak relayed the connection to the Chicago FBI agents working the case today." SA Goodman began but was interrupted abruptly.

"Andrew was not my son." Bryce Stanley corrected SA Goodman. Jesse Goodman didn't know how to respond. Deklan Novak stared at Bryce Stanley. "Your half-brother was not my son. I do not see how that case is relevant today. The boys were lost more than two decades ago." Stanley repeated the statement and was aware of Deklan Novak's well-publicized path from Navy SEAL to local football hero to the FBI. "Your mother's school will be held responsible for allowing my grandchildren to be kidnapped. I cannot imagine the school will survive much longer." Bryce Stanley was pointing at Julia Novak.

"Mr. Stanley. Let's focus on getting the children back today." Deklan remarked without allowing the anger he felt running through his veins to show up in his tone or his demeanor. "I know the circumstances that brought us here are beyond stressful and the FBI will do everything possible to locate and return the children safely to your family, but do not threaten my mother again. The Starlight Academy and my mother were not responsible for the circumstances today that have brought us here, Mr. Stanley." Deklan paused and waited for his team to give him their undivided attention.

"You are responsible for what has apparently happened to your grandchildren." Deklan addressed Stanley directly. "The FBI was able to unseal the settlement from the Paige Novak Greene/Bianco Amato lawsuit against North Suburban University Health Systems and the United Network of Organ Transplants in 1998, brought by your firm, the Stanley Law Group. The case was settled out of court. The actual settlement was much higher than your firm reported to the plaintiffs, my father and Ms. Greene's family. Your firm had the plaintiffs sign their settlement agreement for the agreed award of $3 million. Unbeknownst to the plaintiffs, NSUHS and UNOT agreed to pay $43 million, in large part to quiet the case and avoid an avalanche of follow-up lawsuits by waiting organ transplant patients. The money was placed in your firm's account, but the plaintiffs were told that the amount was $3 million not $43 million. The plaintiffs were told that the $3 million had to be doled out in small monthly installments while the remaining portion of the bogus $3 million award landed in an untouchable annuity that paid minimal monthly interest payments. Your firm informed the clients that the judge allocated the $3 million settlement. A lie. You told the clients that the judge had to sign off on everything and every payment, another lie. You told your clients that various medical debts and liens for Paige Novak Greene and a slew of inflated legal fees for Ms Greene had to be paid first, another series of lies. The plaintiffs received less than $90,000 of the $43 million settlement."

"Those documents are protected and sealed." Bryce Stanley blurted out without thinking.

"Sealed documents can be unsealed if a compelling public interest outweighs the continued confidentiality. In this case, I started the process to unseal the records after I joined the FBI and was assigned to the Chicago office. The compelling public interest was solving a cold case that involved the missing boys from New Trier High school. Conveniently for the Bureau and most inconveniently for you, Mr. Stanley, the records were unsealed less than a month ago." Deklan explained.

"Let's tackle the assumption that the Outfit doesn't forgive or forget, no matter how many years go by. The Outfit watched you profit tremendously over the years while Bianco Amato died during the civil trial. The Outfit murdered three young men in 2000 because of a lawsuit that triggered a chain of events ending with Bianco Amato's death. If the FBI can unseal a sealed court document legally, how long do you think it took organized crime to unseal the settlement agreement that awarded you and your firm $43 million? How many federal court employees were bribed or threatened to unseal the settlement agreement back when it was sealed?" Deklan asked rhetorically. "Mr. Stanley, you are solely responsible for the kidnapping of your own two grandchildren." Novak was uncomfortably close to Bryce Stanley's space, and no one was about to interrupt the former Navy SEAL and current Special Agent of the FBI, except his immediate colleague supervisor. Special Agent Jesse Goodman pulled aside Special Agent Novak and whispered something in Deklan's ear.

"The Wilmette Police Department just found the bodies of Hillary Stanley and Brittany Hawn in the basement of the Stanley home. Both women were shot in the head with a Colt M1911 .45. The gun was left at the scene.

The story will continue in a second Deklan Novak novel…

ABOUT THE AUTHORS

TOM HRUBY

Tom graduated from Andrean High School in Merrillville, Indiana in 2000. Tom played Varsity football as linebacker, fullback, and D-End in high school. Tom was MVP on the wrestling team during HS Senior year. After foregoing college, Tom became a professional MMA fighter in Chicago going undefeated at 7-0. After 9/11, Tom worked construction while he trained for the Navy and the grueling Navy SEAL program. Tom entered Navy bootcamp and BUD/S training in 2006. Tom graduated from BUD/S, Airborne School and SQT, receiving his SEAL Trident in 2008. Tom joined SEAL Team One as a Breacher in 2008 and served until 2016 with four deployments including Iraq and Afghanistan. Tom worked closely with DEVGRU, ISOC, CIA, DEA, FBI and other government agencies tracking and capturing terrorists globally, culminating in the capture/kill of Osama Bin Laden in Pakistan 2011. Tom left Afghanistan one month before the raid in Abbottabad. While still an active-duty Navy SEAL Pre-BUD/S Instructor (Lead Petty Officer LPO) stationed at Great Lakes Naval Recruit Training Command (RTC Great Lakes) in North Chicago, Illinois, Tom enrolled at Northwestern University as a full-time student and walked onto the football team as a linebacker after not playing football for 15 years. Tom played in the Big Ten for Pat Fitzgerald and the NU Wildcats from 2014-2016 at 32, 33, and 34 years old. Loving partner and the father of five boys, Tom currently lives at Hruby Ranch in Crown Point, Indiana where he coaches lots of football, wrestling and baseball.

JAMES POMERANTZ

James studied Journalism at the University of Missouri. He wrote his first novel in 1999 and cut his writing teeth in Chicago while owning 10 nightclubs and sports bars over many years. James owned a restaurant early in his career sponsored by the Mario and Michael Andretti IndyCar racing team owned by actor Paul Newman and Carl Haas. Andretti's 1987 Indy car was in the main dining room. Also, James survived a restaurant partnership with martial arts action-star Steven Seagal during Seagal's glory days following the breakout hit movie, *Above the Law*. James described the business relationship with Seagal as interesting. "It was before Seagal went to the other side." James Pomerantz novels before The Breacher's Playbook include Ghost Bandit, Undisclosed Sibling and The Spitting Image of My Father. James wrote a 2004 biography entitled They Call Me Sid Rock… Rodeo's Extreme Cowboy. (Triumph Books, Chicago) The book was a colorful expose of Sid Steiner, 2002 PRCA World Champion Steer Wrestler. The Steiner family is Texas rodeo royalty. Sid's father Bobby Steiner was a PRCA World Champion Bull Rider in 1973. James lost his wife, Mary, in 2019. They raised four kids in Winnetka, Illinois. James now resides in the Chicago area near his family.

ACKNOWLEDGMENTS

- Kathy
- Thomas Troy III
- Ethan
- Dean
- Beau
- Reed
- Brooke
- Tom Sr.
- Megan
- Milos
- Mary Courtney Gilbert Pomerantz
- Jimmy, Kiley, Michael, Matthew, Erin, Christin, Mary Brock, Robbie
- Linda Pinsky-Wolkoff
- Master Chief John Clift: Naval Amphibious Base Coronado, United States Navy SEAL
- E-6 Gene Mak...United States Navy SEAL
- Chief Hawk Slater...United States Navy SEAL, BUD/S training 1988, ST4 1989-1997, ST1 2002, ST6 2005-2010, ST1 2012-2015
- Pat Fitzgerald...Northwestern University Head Football Coach 2006-2022
- Randy Bates...Northwestern University linebacker coach 2006-2017, Pittsburgh defensive coordinator 2018-present
- Dan Vitale...Northwestern University fullback 2012-2016, NFL 2016-2020, Bucs, Bills, Browns, Packers, Patriots

- John Lauer: Special Agent FBI Quantico Training Academy, FBI Hostage Rescue Team, Lead Instructor Hogan's Alley, Navy SEAL 1998-2005 ST4 & ST8
- Raymond Hall: Special Agent FBI Washington D.C.
- Joan Froehle: FBI Academy Special Events Team Supervisory Management, Quantico, VA.
- Jenny Nichols: Unit Chief Coordination and Support Unit Federal Bureau of Investigation FBI-Training Academy
- Retired FBI Special Agent Ken Marischen…1966-1993, San Diego, Los Angeles, and Anchorage
- Dr. Richard Pomerantz: retired general surgeon, Section Head, General Surgery at St. Joseph Mercy Hospital, Medical Director, Trauma Center and Medical Director, Surgical Intensive Care Unit at St. Joseph Mercy Hospital in Ann Arbor, Michigan
- Tom Johnston…Tom Hruby's high school football coach
- James Howser: Steamboat Sotheby's International Realty Steamboat Springs, Co.
- Jim Morrissey: Chicago Bears linebacker 1985-1993

About the Publisher
TACTICAL 16

Tactical 16 Publishing is an unconventional publisher that understands the therapeutic value inherent in writing. We help veterans, first responders, and their families and friends to tell their stories using their words.

We are on a mission to capture the history of America's heroes: stories about sacrifices during chaos, humor amid tragedy, and victories learned from experiences not readily recreated — real stories from real people.

Tactical 16 has published books in leadership, business, fiction, and children's genres. We produce all types of works, from self-help to memoirs that preserve unique stories not yet told.

You don't have to be a polished author to join our ranks. If you can write with passion and be unapologetic, we want to talk. Go to Tactical16.com to contact us and to learn more.

All of Tactical 16's books are available on our online bookstore, T16Books.com. Visit it today to see more books from our selection of authors and to find a new adventure to read!

www.ingramcontent.com/pod-product-compliance
Lightning Source LLC
Chambersburg PA
CBHW050856240426
43673CB00008B/261